Why Can't a Woman Be More Like a Man?

The Evolution of Sex and Gender

LEWIS WOLPERT

FABER & FABER

First published in 2014
by Faber and Faber Limited
Bloomsbury House
74–77 Great Russell Street
London WC1B 3DA

Typeset by Palindrome
Printed and bound by CPI Group (UK) LTD, Croydon, CRO 4YY

A CIP record for this book
is available from the British Library

ISBN 978-0-571-27925-8

2 4 6 8 10 9 7 5 3 1

Contents

Acknowledgements

I am grateful for the advice given to me by my editors Julian Loose and Kate Murray-Browne at Faber, and my agent Anne Engel. Dr J. Herberg and Professor C. Tickle did valuable editing. My partner Alison Hawkes made a major contribution by helping, for many hours, to revise the final version of the book, and I could not have completed it without her assistance. Paula Turner made a valuable contribution by editing the manuscript and Sarah Barlow in reading the proofs.

I

Questions

> The female is softer in disposition, is more
> mischievous, less simple, more impulsive and more
> attentive to the nurture of the young. The male, on
> the other hand, is more spirited than the female,
> more savage, more simple and less cunning.
>
> Aristotle

My title is from the famous plaint by Professor Higgins in the musical *My Fair Lady*. The song is about the difference between the sexes. It comes very clearly from an amusingly, ludicrously biased male point of view, but it serves to remind us that the differences between men and women are a major issue in our lives. However, I might just as well have called the book *Why Can't a Man be More Like a Woman?* Since men, as I shall show, are essentially biologically modified women.

So just how different are men from women, and are there important differences that are genetically determined? This is a very controversial subject, but most men and most women would think that there are significant differences, and so do I. An enormous scientific literature, often disputed, has been devoted to the subject and it is one that arouses strong passions. There has even been criticism of some of this scientific literature to add to the controversy – namely, the assertion that some investigators may have biased their research deliberately to show differences, since a result of an investigation showing no difference between males and females would be less interesting.

Are there significant biological factors that make women behave differently from men, or are all the apparent differences due to social and cultural factors? I am a developmental biologist who has studied how embryos develop from the fertilised egg. Genes control the development of the embryo by providing the codes for making proteins which largely determine how cells behave. The cells in the human embryo give rise to the structure and function of our brains and bodies. We are essentially a society of cells which come from the egg and which determine who we are. They determine whether we are male or female, and I want to understand whether important differences in the behaviour of men and women are largely controlled by their genes during development and by the action of hormones both in the womb and in later life. This is particularly relevant in relation to men and women's intellectual, emotional and physical abilities and their social behaviour.

And why are there two sexes? This is a difficult problem related to our evolution, and most animals have two sexes. But there are some species with just one, and with these, asexual reproduction is possible, if simpler and less enjoyable. Evolution has selected differences between men and women so as to make their reproduction as successful as possible. What are the evolutionary advantages of two sexes and of the genetically specified differences between men and women? Are there differences in intelligence and mental skills, language and motor skills?

The development of the brain, which determines how we behave, involves billions of nerve cells making innumerable connections with each other, and this is controlled by genes. We need to understand the differences that develop between male and female brains. The hormone environment in the embryo, controlled by genes, clearly contributes to the development of the physical and mental differences between males and females. Genes on the sex chromosomes explain why most people are heterosexual, for example. In other animal species, research has shown that the early hormone environment clearly has long-term effects on behaviour by changing the development of the brain, but the specific changes involved in humans have only recently begun to be identified and are a controversial area.

Also controversial are the many sex differences that have been described between human male and female brains, but only a few of these seem to be relevant to sex differences in behaviour. Regions of the brain, such as the hypothalamus and amygdala, which play key roles in processing emotions and sexual behaviour, have receptors for male hormones like testosterone, but much work still needs to be done in discovering reliable links between these hormones, the development of the brain and different behaviour in men and women. I will look at the structural differences between male and female brains revealed by the powerful techniques of MRI (magnetic resonance imaging) as well as by post-mortem examinations. MRI allows one to identify which bit of the brain is active under different conditions, such as thinking about a particular problem, being under stress or engaging in sexual activity.

One well-supported scientific view is that there are inborn differences between the minds of men and women. But this view is being challenged by scientists who call this the pseudoscience of 'neurosexism', and are raising concerns about its implications. They emphasise instead social influences, such as stereotyping, in determining the differences in the behaviour of the two sexes. If, for example, parents tell boys that they have less chance of acquiring good verbal skills than their sisters, and tell girls that they have little prospect of developing skills in maths, this can seriously and negatively affect their acquisition of skills in these subjects.

The evidence for some genetic differences in males and females is, however, overwhelming – just consider our bodies. Yet it has been cogently argued that sex differences in behaviour do not have a genetic basis, instead being socially determined. In this book I shall question the claims made for greater ability of males or females in a number of intellectual and emotional spheres, looking in detail at maths and science, motor abilities, analysis of complex systems, empathy and language skills. Much of the recent literature claims that men and women differ in a number of emotions, linguistic ability, memory, spatial reasoning and even in their sense of humour. If the differences are significant it could mean that men and women are naturally suited to different kinds of work. For example, the small number of women employed in areas related to engineering and technology might be related to their psychological function.

Some commentators see differences to be solely a matter of biology, with no social or cultural interventions, though, as we shall see, social conditioning can shape some biological features. The *Oxford English Dictionary* defines feminism as 'Advocacy of equality of the sexes and the establishment of the political, social, and economic rights of the female sex', and some feminists promote what is called the similarities hypothesis, which contends that the psychological similarities between males and females outweigh the differences. In effect, males and females are more alike than they are different.

All this will be considered in some detail, as will differences in emotional attitudes between men and women and the nature of sexual attraction.

In discussing biological differences I have, with just a few exceptions, left out the enormous literature on other male and female mammals such as chimpanzees. There is some hesitation in accepting its direct relevance to humans and, while I think many studies are relevant, many are not. In addition, I rarely refer to behavioural and brain studies on mice – we like to think we are very distinct from them.

Helen Bradford Thompson made the earliest scientific study of sex differences in cognitive abilities in 1903 at the University of Chicago. Her pioneering work failed to show any differences in the emotional abilities of men and women, and only very small differences in intellectual capabilities, which she attributed to social conditioning. But early brain studies concluded that women were intellectually inferior simply because they had smaller and lighter brains. Since then there has been a great deal of research, but questions still remain. In an interview for the *New Scientist* to mark his seventieth birthday in January 2012, the theoretical physicist Stephen Hawking admitted he spent most of the day thinking about women. 'They are', he said, 'a complete mystery.'

In discussing male–female differences the term 'gender' is not used in the same way as 'sex'. Gender refers to socially constructed roles and characteristics used to distinguish between males and females in a given society. There is a distinction between biological sex and gender, as sex refers to the biological characteristics, namely the genetic constitution of an individual, whereas gender refers to their personal identity, beliefs and behaviour. Sexologist John Money first made the distinction between the terms almost sixty years ago. The word 'gender' grew to be used in relation to how individuals identify themselves. The words 'male' and 'female' are sex categories, while 'masculine' and 'feminine' are gender categories, according to the World Health Organization. Differences between men and women, as I have said, are controversial and there are many books on the subject, some of them science-based. But there are also very popular books like John Gray's *Men Are from Mars and Women Are from Venus*, and *Why Men Don't Listen and Women Can't Read Maps* by Allan and Barbara Pease. These amusing books focus largely on widely held beliefs about differences in communication and the relationships of couples, and stress the possible innate differences between the minds of men and women. There are, of course, numerous scholarly books on this topic too, as we shall see.

It has been suggested by Deborah Cameron in her book *The Myth of Mars and Venus* 'that some writers on this subject can be thought of as latter-day Galileos, braving the wrath of the "political correctness" lobby by daring to challenge the feminist orthodoxy which denies that men and women are by nature profoundly different.' Cameron recounts how Simon Baron-Cohen, in his book *The Essential Difference*, explains in his introduction that he put the book aside for several years because 'the topic was just too politically sensitive'. In the chapter in his book about human nature, *The Blank Slate*, Steven Pinker congratulates himself on having the courage to say what has long been 'unsayable in polite company about male–female differences'. Cameron points out that neither writer has a political axe to grind, and that they are simply following the evidence where it leads and trying to put scientific facts in place of dogma. I shall be trying to do exactly that myself and am certainly no Galileo. But my most basic and potentially controversial contribution to the subject is my conviction, based on embryonic development, that men are fundamentally modified females. Not 'Why can't a woman be more like a man?' but rather 'Why isn't a man more like a woman?'

I first look at the evidence for discrimination against women and their subordination throughout history.

2

Discrimination

For most of history, Anonymous was a woman.
Virginia Woolf

Men have dominated all societies from the earliest known times. In almost every society, the primary tasks of men and women have been different. Men are given responsibility for the government of the society, and women responsibility for the daily care of the household and children. Thus the female has been in many ways subordinate to the male and has generally had a lower status. Almost all inventors of modern technology have been male, and similarly most artists who have shaped our world have been men. There have, of course, been some very important women working in these areas, but historically the list is tiny in comparison to the number of men. What is the basis for this? Is it social or biological?

Understanding the differences between men and women could help to explain why women have suffered so much discrimination. Men and women are certainly different, as their sexual behaviour makes all too clear, and all societies accept this, but the belief that women are biologically and intellectually inferior to men has also been widely held and has an ancient history. Hunting and gathering was the ancient means of getting food, and men did the hunting while women gathered food locally and looked after the children. Women were there to provide domestic comfort and, of course, children, but many were looked on more as servants than equals.

In a very small number of societies women have played a more powerful part. One example is the Iroquois in North America, in whose society mothers have important moral and political roles. Another example is in Meghalaya in north-eastern India, where women run businesses, dominate households and take all key family decisions. They also inherit all property. There are other matrilineal societies where descent is via the female line. It is also claimed that in primitive societies this system preceded that of descent through males.

Women were subordinate in the most ancient cultures, such as those of the ancient Mesopotamian societies of Sumer and Assyria, whose religions and laws even prevented women from having control over their reproductive function. In ancient Egypt, however, women were given equal rights with men under the law. They could initiate divorce, control their own property and finances, and appear in court as witnesses. A woman could even become a pharaoh, though only a handful did. Even so, men still played a far larger part in government while women continued to spend most of their time at home. Nevertheless the Greek historian Herodotus, who claimed he had visited Egypt around 450 BC, thought the Egyptians had 'reversed the ordinary practices of mankind'. In the ancient civilisation of Babylon women had similar status, enjoying complete independence and equal rights with their husbands and brothers.

In Greek society the status of women was determined mainly by the standing of the men in their lives, their husbands and fathers. In Athens wives seem to have been considered of use only in producing children, particularly male children, and contributed little to society in their own right. Around 800 BC the poet Hesiod wrote about how Zeus, chief of the gods in Greek mythology, created a beautiful first woman, Pandora, who was moulded out of earth. She was given unique gifts including a jar not to be opened in any circumstances. But she was tempted to open the jar that she had been given and released terrible evil into the world. There are striking similarities between the legend of Pandora and that of Eve in the Bible, as both are about the creation of the first woman, her disobedience of a divine command, and her thereby becoming a disseminator of evil, a status which was extended to all females. Eve ate an apple from the Tree of Knowledge and she and Adam were accordingly expelled from the Garden of Eden. Women were already doing badly in these accounts – which came, of course, from men.

Aristotle believed females to be physiologically and psychologically inferior to males. Yet in Greek mythology the Amazons were a nation of female warriors, and in classical Greek and Roman religions goddesses were worshipped. But their apparent high status did not help their earthly sisters much. With respect to reproduction, a woman was seen as essentially an infertile male, and the development of the embryo was attributed entirely to the male's input of sperm.

In early Roman law women were inferior to men and were described as children. Although Roman women were citizens, they could neither vote, run for office nor work in government. Young unmarried women were controlled by their fathers, and on marrying passed into the control of their husbands. Later, under the Roman Empire, women were give more freedom and had more public influence, even if this was largely limited to those in the educated upper classes. But women in general were still dependent, mistrusted, and looked upon as frail creatures.

The early Christian church adopted the Greek philosophy that held women to be inferior to men by nature and based its laws on the Roman legal codes which gave women fewer rights than men in the home and in civic society. Women's presumed inferior status was linked to scriptural texts which claimed that only man was created in God's image. Just consider: 'But I would have you know that the head of every man is Christ; and the head of the woman is the man; and the head of Christ is God' (1 Corinthians 11:3), and 'Of the woman came the beginning of sin, and through her we all die' (Ecclesiasticus 25:24).

Christianity has traditionally given men the positions of authority, while few women are mentioned in the Bible by name and role, and they are seriously downgraded. 'Wives, submit yourselves unto your own husbands, as unto the Lord. For the husband is the head of the wife, even as Christ is the head of the church: and he is the saviour of the body . . .' (Ephesians 5:22–3); 'Let your women keep silence in the churches: for it is not permitted unto them to speak; but they are commanded to be under obedience, as also saith the law' (1 Corinthians 14:34); 'Unto the woman he said, I will greatly multiply thy sorrow and thy conception; in sorrow thou shalt bring forth children; and thy desire shall be to thy husband, and he shall rule over thee' (Genesis 3:16).

Later Christian theologians and authors continued with these negative views. Tertullian, born in the second century AD, described women as 'the devil's gateway' in his *De Cultu Feminarum* and criticised their love of gaudy clothing. Fathers of the Christian Church such as St Augustine believed that only men, not women, were made in the image of God and St Jerome was another vehement misogynist who considered women to be the root of all evil and said, 'Woman is the gate of the devil, the way of wickedness, the sting of the serpent, in a word a dangerous thing.' These doctrines influenced Christian theology for centuries. Thomas Aquinas still maintained in the thirteenth century that women were intellectually inferior and created only to help men by conceiving children.

Women in ancient Chinese culture had virtually no rights. According to Confucius, women did not deserve an education because they were not equal to men. It was the duty of a Chinese girl to obey the men in her family, and have her feet bound to make her more desirable for marriage, and sometimes not even be given a name but be known as 'daughter number one' or 'daughter number two'. In ancient India, by contrast, women were originally valued equally to men, but this positive view gradually declined as females came to be deemed less useful or valuable than males.

At the time of the foundation of Islam women had little or no rights. But the prophet Muhammad revolutionised this by giving women rights including property ownership, inheritance, education and divorce. The Koran is clear that men and women are equally valuable, although men have superior abilities which make it their duty to protect women. However, even in recent times women under the extreme Islamist regime of Taliban-controlled Afghanistan were forced to wear the burqa in public and forbidden to work, while girls were denied education after the age of eight.

The origins of the movement for women's suffrage are attributed to late eighteenth-century France, but it was not until 1839 that Mississippi became the first of the United States to pass laws allowing married women to own property separately from their husbands; other states followed. It was only in 1893 that the British colony of New Zealand became the first self-governing nation to extend the right to vote to all adult women. The first European country to introduce women's suffrage was the Grand Principality of Finland, which elected the world's first female members of parliament in 1907. In 1928 British women won full voting rights on the same conditions as men. But even now women worldwide have limited political power. In the parliament of the United Kingdom less than twenty-five per cent of members of both Houses of Parliament are women at the time of writing, and in the United States Congress the figure is only eighteen per cent.

In practically all economically primitive societies men perform physically demanding work such as hunting, fishing, metalworking, weapon-making and boat building. The women normally farm, manufacture and repair clothes, and do the work at home, which, as we shall see, has led them to make important inventions in agriculture. The same division of labour has been observed all over the world. There are a few examples of societies, such as the aborigines of the Trobriand Islands, where the system of inheritance is matrilineal and women hold a very good position, but men remain in charge. Even in most modern societies, men are more likely to do physical work and travel, while women more commonly stay at home and take responsibility for domestic work and caring for children.

A special form of discrimination is female genital mutilation, involving partial or total excision of the female genitals for non-medical reasons. It is thought to affect some 140 million women worldwide. The prevalence is particularly high in north-east Africa. The reasons vary, but a key one is 'purification' of the woman, and even some mothers support it. It is a form of gender-based violence.

Is the domination of women by men dependent on the different biological features of men and women? To explain the basis of this historical and current situation, it has been often set forth as indisputable that men are socially superior because they are naturally superior. Men, many have claimed, are endowed by nature with higher physical and mental attributes. Consider Charles Darwin's views. In *The Descent of Man* (1871) he wrote:

Man is more courageous, pugnacious and energetic than woman, and has a more inventive genius . . . Woman seems to differ from man in mental disposition, chiefly in her greater tenderness and less selfishness; and this holds good even with savages, as shewn by a well-known passage in Mungo Park's *Travels*, and by statements made by many other travellers. Woman, owing to her maternal instincts, displays these qualities towards her infants in an eminent degree; therefore it is likely that she would often extend them towards her fellow-creatures. Man is the rival of other men; he delights in competition, and this leads to ambition which passes too easily into selfishness. These latter qualities seem to be his natural and unfortunate birthright. It is generally admitted that with woman the powers of intuition, of rapid perception, and perhaps of imitation, are more strongly marked than in man; but some, at least, of these faculties are characteristic of the lower races, and therefore of a past and lower state of civilisation.

The chief distinction in the intellectual powers of the two sexes is shewn by man's attaining to a higher eminence, in whatever he takes up, than can woman – whether requiring deep thought, reason, or imagination, or merely the use of the senses and hands. If two lists were made of the most eminent men and women in poetry, painting, sculpture, music (inclusive both of composition and performance), history, science, and philosophy, with half-a-dozen names under each subject, the two lists would not bear comparison. We may also infer, from the law of the deviation from averages, so well-illustrated by Mr. Galton, in his work on Hereditary Genius, that if men are capable of a decided pre-eminence over women in many subjects, the average of mental power in man must be above that of woman.

But Darwin had no real evidence for that conclusion.

Biological differences have often been blamed for the inequality between the sexes. Nicolas Malebranche, a French religious philosopher, argued in the seventeenth century that abstract thought was impossible for women because of the delicacy of their brain fibres. In 1875 the English biologist and philosopher Herbert Spencer, who coined the term 'survival of the fittest', argued that women were incapable of abstract thought and could not understand questions of justice, only issues of care. Sigmund Freud described women as inferior to men and argued that they were deficient in abstract thought, more influenced by feeling than reason. 'Women oppose change, receive passively, and add nothing of their own,' he wrote in 1925. Freud also proposed that girls and women were incomplete men. His concept of penis envy was based on the notion that girls realise early that they are not as complete as boys, and want to be more like them. I cannot take Freud seriously in spite of his great influence.

In contrast to all the negative views about women, fiction has been much more positive. Virginia Woolf in *A Room of One's Own* claimed:

> Indeed, if woman had no existence save in the fiction written by men, one would imagine her to be a person of the utmost importance; very various; heroic and mean; splendid and sordid; infinitely beautiful and hideous in the extreme; as great as man, some think even greater.

It takes much research to understand and account for men and women's current unequal outcomes in academic careers. There are fewer women in academic positions in universities, especially in maths and science. A recent survey in the United States showed that women in science, engineering and technology were much less likely to obtain tenure in full-time academic positions than men. The figures were twenty-nine per cent of women compared to fifty-eight per cent of men. In addition only twenty-three per cent of women compared to fifty per cent of men achieved the rank of full professor. Why are women so under-represented in this sphere? What role do biological and social factors play? We will return to this later.

It is important to recognise how much the position of women in modern society has changed. Women have developed feminism, a movement based on the belief that women should have equal rights to men in all spheres of life, with widespread success. Many women hold high positions in politics as well as the commercial world; consider Angela Merkel in Germany and Hillary Clinton in the United States, and remember Margaret Thatcher. There are several awards in the United Kingdom for achievements by women in business. In the Olympic Games women have an equal role to that of men. Yet many people still consider housekeeping and raising children to be a woman's proper sphere. The feminist Germaine Greer claims that there has been a recent rise in misogyny as women expect to share men's lives and men find this intolerable.

Do these widely observed differences in male and female social roles arise from biological foundations? We return to the tricky set of problems of identifying human male–female differences and then determining whether these have a social or biological cause. I will first look at the embryonic development of humans to see how the physical differences between the sexes have arisen.

3

Modified Women

What would men be without women? Scarce, sir
. . . mighty scarce.

Mark Twain

The biological differences that can be found between the bodies and brains of males and females are largely due to the way their embryos develop. After starting life in the womb as a single cell, the fertilised egg, we all begin to develop as females, the default sex. Early development of the human embryo follows a female path and is thus similar in males and females, with sexual differences appearing only at later stages when male development starts by modifying female development. Although the development of the individual as either male or female is genetically fixed at fertilisation, males develop only because, in about half the embryos, there are genes that modify female development. If these genes did not do their job properly, we would all be women. A sobering thought for us males.

The fertilised egg divides to give rise to all the cells in the body, and the organisation of the developing embryo is determined by the behaviour of cells in different regions. Cell behaviour is largely determined by their proteins, specified in turn by genes that provide the codes for the different proteins. In humans some 23,000 genes are located on forty-six chromosomes, twenty-three from each parent. Turning different genes on and off in different places and at different times in the developing embryo determines what proteins the cells contain, and thus how the cells behave. In this way genes play a fundamental role in how we develop. Genes, via the proteins they code for, determine the shape and function not only of the body but also of the brain, so that certain behaviours are genetically determined. An obvious one is the desire to have sex.

Whether a gene is switched on, and thus whether its protein is made, depends on the binding of certain proteins – transcription factors – to the DNA in special control regions of the gene. Only if the correct control regions are occupied by the right transcription factors can a gene be turned on. The activation of a single gene may involve many transcription factors, and they can also turn genes off so that the proteins they code for cannot be made. In addition hormones can affect the control regions by regulating the release of transcription factors or preventing them from acting. The protein produced by one gene can activate several other genes or inactivate them, and so a circuit of gene interactions is set up which determines cell behaviour and how it changes with time.

The basic mechanisms involved in embryonic development are understood, but the role of many proteins remains to be deciphered, and this is especially true for the development of the brain. A key question is what controls the behaviour of individual cells so that highly organised patterns emerge, like those of the nerve cells in the brain. The cells arising from division of the fertilised egg become different from each other as a result of signals between each other as well as of their internal developmental programmes. Hormones released by cells in one part of the embryo can affect cells in other parts, so that they too play a key role.

Signals from other cells often arrive at the cell membrane but most do not enter the cell, instead binding to a receptor which activates a sequence of molecular interactions within the cell leading to genes being turned on or off. This sequence can be very complex and there is a nice cartoon by Rube Goldberg, *The Self-Opening Umbrella*, which illustrates it. A man has a chain of interactions that causes his umbrella to go up when it rains. The rain causes a prune to expand and so light a lighter which starts a fire which boils a kettle which whistles and frightens a monkey who jumps on to a swing which cuts a cord releasing birds who, when they fly out, raise the umbrella. In the cell the sequence can be much more complex.

After fertilisation, the egg undergoes a number of cell divisions which give rise to a layer of cells that will develop into a human body. This layer then undergoes gastrulation, a sequence of movements which sets up the basic structure of the embryo (I am quoted as saying that gastrulation, not birth, marriage or death, is the most important event in our lives). After gastrulation the nervous system begins to develop, starting with the formation of the neural tube from a sheet of cells. The anterior end of this tube gives rise to the brain, while further back it develops into the spinal cord. The nervous system is the most complex of all our organ systems, as there are many hundreds of types of neuron, differing in the billions of connections they make. Some features of its development will be described later.

It is the female that dominates in embryonic development. She provides the cell – the egg that gives rise to the child – while the male contributes only his genes at fertilisation. She also feeds and protects the embryo during its development. Early development of the human embryo, as we have seen, follows a female path. But the development of the individual as either male or female is genetically fixed at fertilisation by the chromosomal content of the sperm that fuses with the egg. There are two chromosomes that determine the sex of an embryo, X and Y. Females have cells with two X chromosomes (XX) while males have an X and a Y (XY). It is worth noting that in women one of the two X chromosomes is randomly inactivated, as only one is required for normal function. Before fertilisation the cells that give rise to egg and sperm divide and halve the number of their chromosomes, the beginning of a process known as meiosis. The egg thus has a single X chromosome while each sperm cell carries either an X or a Y. The genetic sex of mammals like us is thus established at the moment of conception, when the sperm introduces either an X or a Y chromosome into the egg. The presence of a Y chromosome makes the embryo develop as a male; in its absence, the default development is along the female pathway. If the cells in the embryo have just a single X chromosome and no Y – Turner syndrome – the embryo develops as female, but with abnormalities. In the absence of the X chromosome, there is no development.

The genes on the Y chromosome carry code for only about twenty proteins, but a most important key gene is SRY, which causes testes to develop. The testes then secrete hormones about ten weeks into gestation including testosterone, which cause the development of male tissues and suppress female development. Testosterone is high in male embryos during weeks twelve to eighteen and then again during weeks thirty-four to forty-one, when it reaches a level ten times higher than that found in female embryos. It is also high in baby boys in the first three months after birth. The development of a penis in males, instead of the clitoris of females, and the reduced size of mammary glands in males are due to the action of testosterone. Breasts and nipples are already in place in the embryo before testosterone is secreted, as it is essentially female at early stages, so they are present in both sexes but have no function in males. Thus the SRY gene has a major effect on our lives: it initiates the basic differences between men and women by turning on male hormones. In the male embryo the SRY gene is expressed in the brain, kidneys, heart and pancreas, while in adults its activity can be detected in the kidneys, heart and liver. It is very clear from our sexual development that males are essentially modified females, and that applies to both our bodies and our brains.

Imprinting is a peculiar and special feature in development. During maturation of the sperm and egg certain genes are inactivated by a process known as imprinting. At least a hundred genes are affected; the majority of their functions remain to be worked out, although most are expressed in the brain. Their general function is to control development so as to give an advantage to whichever sex is involved in the imprinting. Thus imprinting in the egg will reduce the growth of the embryo in order to limit the negative effects excessive growth can have on the mother. By contrast male imprinting of the sperm promotes growth. It is clear that there are significant genetic differences in male and female brains due to imprinting.

There is a higher rate of left-handedness among men than among women, and this may be due to the higher levels of prenatal testosterone in males. A minor physical characteristic that shows sex differences is the ratio of the length of the right-hand second digit (index finger) to that of the fourth digit (ring finger). In men the second digit tends to be shorter than the fourth, while in women the second tends to be the same size or slightly longer than the fourth. Lower 2D:4D ratios imply a higher testosterone exposure in the embryo.

The effect of hormones such as testosterone on the development of male features of the embryo is fundamental to understanding biological differences between the sexes, including physical characteristics such as size and strength as well as particular skills and emotions. More than fifty years ago a paper was published about the effect of treating the embryos of female guinea pigs with male hormone, which masculinised their sexual behaviour. Since then there have been numerous studies showing that sex hormones present early in development affect sexual differentiation and sex-specific behaviour. It is not permissible to inject male hormones into the human womb where female embryos are developing to observe the effect, but it is possible, as we shall see, to study the development of individuals with disorders that cause an increase in testosterone in the womb, as well as measuring hormone concentration during pregnancy.

Hormones are not the sole contributor to the differences in typical male and female development. Genes on the sex chromosomes influence the development of both the body and the brain, and directly influence neural developmental and some identifiably sex-specific behaviour. These direct genetic actions have a wide influence and can include actions caused by locally produced hormones or other, non-hormonal, messenger molecules.

One informative disorder of sexual development

occurred on the island of Hispaniola in the Caribbean, where children who had been identified as girls at birth began to develop male sexual characteristics at puberty, and grew up to become men. This was found to be due to a faulty gene for the enzyme 5-alpha reductase, which is required to make testosterone potent. Its absence in embryonic development was resulting in babies being born with female characteristics even though they had a Y chromosome, but at puberty the secretion of large amounts of testosterone caused them to grow a penis. The defect was traced back for seven generations. We will also examine examples of another genetic disease, congenital adrenal hyperplasia (CAH), which increases testosterone during development and so masculinises the development of the female embryo, and causes changes in the female body and later behaviour.

Further evidence for the key role of genes and hormones in human sexual development is illustrated by rare cases of abnormal development. Testosterone can enter cells and then bind to a receptor. If XY males have a mutation that prevents testosterone binding, they develop as females in external appearance, even though they have testes that secrete testosterone, and they do not develop male secondary sexual characteristics at puberty. Thus there are males with this insensitivity who despite their Y chromosome are usually indistinguishable from girls and women. Conversely, genetic females with a completely normal XX constitution can develop as male in external appearance if they are exposed to male hormones during their embryonic development. There are also rare cases of XY individuals who are female due to part of the Y chromosome with the SRY gene being lost, and XX individuals who are physically male due to that part of the Y chromosome being transferred to the X chromosome. The sex chromosomes can have negative effects when combined in an abnormal manner that results in there being three of them, instead of the usual two. Some individuals who have three sex chromosomes (either XXX, XXY or XYY) have clearly marked difficulties in speech and language, motor skills and intellectual achievement. Mental ability has been found to be poorest in females with XXX, while males with XYY have normal-range intelligence.

After birth, development continues mainly as growth but at the age of about ten to fourteen years old puberty occurs, bringing major physical changes through which a child's body develops into that of an adult capable of reproduction. Puberty is begun by hormonal signals from the brain to the ovaries and testes, which respond by producing a variety of hormones that stimulate growth. The principal sex hormone for males is testosterone, while the hormone that dominates female development is an oestrogen, estradiol. Females also produce testosterone, which is made in their adrenal gland and ovaries. The concentration in their body is only one twentieth that of males. There is a growth spurt in the first half of puberty which only stops once puberty is complete. Puberty introduces much wider changes into bodies of boys and girls than merely genitalia, and results in radical developments. Females develop functional breasts and oestrogen widens the pelvis to help with childbearing. Males become sexually competent, develop facial hair and their voices break.

The average height of men in the United Kingdom is five feet nine inches, and for women just under five feet four inches. American men are about an inch taller, and American women half an inch. According to the *Guinness Book of Records 2013* the shortest living man – at one foot nine and a half inches – is Chandra Bahadur Dangi of Nepal, and the tallest living man is Sultan Kösen of Turkey, at eight feet three inches. The tallest man ever was Robert Pershing Wadlow (1918–40) from the United States, who was over eight feet eleven inches tall. The tallest living woman is Yao Defen of China at seven feet seven and a half inches, and the shortest Jyoti Amge of India at two feet seven inches.

On average males are physically stronger than females and this has a significant effect on lifestyle, particularly in relation to defence, hunting and aggression. Men have more total muscle mass than women because of higher levels of testosterone. As a result gross measures of body strength suggest that women are between fifty-two and sixty-six per cent as strong as men in the upper body, and between seventy and eighty per cent as strong in the lower body. Males typically have about fifty-six per cent greater lung volume proportional to body mass, and larger tracheae and bronchi. Men also have larger hearts and a ten per cent higher red blood cell count: more haemoglobin, and hence greater oxygen-carrying capacity. An important consequence of these structural differences is that maximal exercise capacity in women may be more limited by lung capacity than in men.

It is worth noting some other physical differences that result from development. Both males and females have an Adam's apple but it is larger in men because their larynx, which contains the vocal cords, is larger. The Adam's apple is actually the thyroid cartilage which forms part of the larynx protruding in the front of the neck. In most women it is not easy to see. Women, who have smaller vocal cords, normally have higher-pitched voices than men. But this is not the only difference. Timbre, the tone quality differentiating one spoken sound from another, also matters and men tend to speak more monotonously while women tend to use a wider range of tones.

Differences are found also in the special senses. In sense of smell, for example, women are significantly better than men in odour detection and identification. It has been claimed that this superiority can even be found in newborn babies. Female sensitivity to male-specific odours is many times stronger during ovulation than during menstruation. Women are also significantly more likely than men to suffer from cacosmia, where a normally benign smell is registered as unpleasant. Symptoms include dizziness, headaches and anxiety, and while there can be more serious causes, links have been found to common household solvents such as paint. Women are better at hearing at high frequencies than men and have even more acute hearing around the time of ovulation. A boy may need a teacher to speak more loudly than girls if he is to follow a class, but he can also block out background noise in the classroom better than girls.

As we shall see from the behaviour of newborn children, there are also psychological differences specified in development as the brains of males and females are different at birth.

One may ask why specific differences should have been favoured and selected for by evolution. And how could they confer advantages for reproduction?

4

Two Sexes

You don't have to be naked to be sexy.
Nicole Kidman

What is the origin of men and women? And why are there
two sexes? An opinion poll in 1999 found that one in two
Americans believed the Bible's account of the origin of
human beings, namely that God made man, and women
came from his rib, so women were clearly a second-class
derivative. Questioning our origins has a long history, but
it was only with Darwin's theory of evolution that any
proper scientific understanding became possible.

It was a common belief in most religions and cultures that
gods had created men and women just as they created earth
and nature. Those early times also fostered other concepts of
human origins. The Greek philosopher Anaximander in the
sixth century BC believed in a gradual development of life
from moisture under the influence of warmth and suggested
that humans originated from animals of another sort. Two
centuries later Aristotle proposed that the world and the
animals and plants living in it were eternal, uncreated and
unchanging. The Chinese creation myth of Pan-ku is one of
the best known of many such stories as to how the world
began. The giant Pan-Ku burst out of a cosmic egg which
had been floating in a formless, chaotic void. Far from
taking just seven days, Pan-Ku spent over 18,000 years
separating the earth and the sky and the opposites of nature,
including men and women. On his death, different parts of

his body became the different features of the earth with his left eye becoming the sun and his right the moon, his blood the rivers, and the fleas on his body scattered to become fish and animals and hence the human race.

Darwin's theory of evolution, first published in 1859, was a fundamental and wonderful change. We humans, he argued, evolved from ape-like animals. The core of his brilliant discovery is that physical and behavioural characteristics can be inherited from one generation to the next, and that these can change so that organisms' offspring can differ from their parents. If this potentially gives them an advantage they will be more successful in reproducing and will increase in numbers more quickly than organisms that lack this inherited advantage. Random change in inherited characteristics, and selection for reproductive advantage, sums it up. The inheritance of characteristics was only later discovered to be entirely due to genes, made of DNA, which are the only structures in the cell that replicate and so can be passed from generation to generation and undergo random change.

Evolution is based on changes in genes which control how the embryo develops. The genetic constitution of an individual is present in the fertilised egg and represents the combination of genes from male and female parents – sperm and egg. Natural selection determines which combinations of genes persist, and in animal evolution this is based fundamentally on successful reproduction. Dogs provide a nice counter-example, as their evolution was based on artificial selection. Dogs were first domesticated from wolves in the Middle East and Europe about 14,000 years ago but in Southeast Asia, the area from where many of our modern breeds were derived, only 7,000 years ago, and recent research from the UC Davis School of Veterinary Medicine says that the dogs of Southeast Asia were on a separate evolutionary track as they could no longer cross-breed with wolves. Breeds from the East may have been the ancestors of many of today's game dogs and were bred to undertake specific tasks such as hunting game. Ancestors of the Afghan hound and the greyhound have been identified at sites in Mesopotamia dating to around about 5000 to 4000 BC, and scent hounds, that hunt by smell rather than by sight and whose descendants include the modern bloodhound, foxhound and dachshund, are recorded from around 3000 BC. Guard dogs for protecting animals as well as people also appeared at around this time. Their descendants were selected by humans for various features, so that they were selectively bred into hundreds of breeds each with its own size, shape and distinctive appearance and abilities. (I am grateful for the evolution of the wire-haired fox terrier, one of which was my childhood friend.) Since humans were intervening to make the selection of animals with specific characteristics, there was very rapid progress, compared with human evolution that had to take its own course: it is estimated that we branched off from the last common ancestor of human and chimpanzee about five to seven million years ago.

Differences in the genes and in their control regions are the raw material on which evolution depends. Evolutionary changes in genes must occur in the germ cells that give rise to the next generation, in our case either sperm or eggs. The evolutionary changes in the genes are expressed either as changes in the nature of proteins themselves, or where and when they are made during the development of the embryo. Those that give rise to the organisms best adapted to their environment with respect to reproduction survive and continue to reproduce – this is the essence of evolution. Thus if genetic changes cause the embryo to develop into an organism that survives better they will be retained in future generations, but if they make things worse, then the changes in the genes will soon be lost.

Basic differences between males and females are thus due to genetic differences that change how the embryo develops and they can affect any structure, including the brain. Such differences will almost always be related to improved reproduction, and so any biological differences between men and women will be understood in terms that explain their reproductive advantage. Of course, social factors can make men and women behave differently and can even alter their biology, and the distinction between biological and social mechanisms is the fundamental and difficult problem that we meet all too often.

One cannot explain the genetic differences between men and women without understanding the role of evolution. Darwin's theory on sexual selection suggests that competition for mates and the careful selection of those who might be suitable as mates have played a role in shaping the evolution of sex differences. It is sex itself that was unquestionably the major factor in determining differences in female and male psychology. Men and women had to feel a need to engage in sexual behaviour, as evolution cares only about successful reproduction. It was during the evolutionary process that their physical and biological characteristics were genetically determined and sex itself evolved. Yet some of the books and articles on this topic do not even mention the word gene, and so totally avoid understanding how genes can determine behaviour.

If women reproduced without men it would make life simpler, and in my view even very boring, without two sexes and the sex act itself. There are indeed animals that do just that, as we shall see. But we, like many animals, are committed to having males co-operate with females in our reproduction.

Females have had the more complex evolution, as they have to provide the egg and nurture the child. Males have only to provide a sperm. It was thus that males evolved as modified females and their evolution was mainly related to protecting and feeding females and offspring. Looking after children and being sensitive to their emotional states and needs could have led to the evolution of empathy, particularly in women; this development would have benefited the children. There is also a view that female behaviour has evolved to help facilitate harmony within the family. Men, meanwhile, needed to carry out tasks which need strength and speed, and the most successful were those who were good hunters and competitive for females. This required aggression, which became genetically determined.

But why this complication? What is the evolutionary advantage of having two sexes? This is a complex issue but it is generally accepted that there are major advantages. Most animals reproduce by having two sexes but there are exceptions, such as the single-sex invertebrates like rotifers. These are very small but their body can be divided into three main sections: a head, a trunk and a foot. For most rotifer species, males have never been discovered; females produce eggs that then develop into females. Virgin birth, in which females produce eggs that develop into young without having been fertilised, also occurs in some insects, fishes, and a few species of lizard and frog.

The evolutionary superiority of two sexes must involve the production of offspring who enjoy some advantage over those created by asexual reproduction. Evolutionary theorists admit that identifying the specific features of sexual reproduction that confer an advantage is a difficult problem in biology. The most favoured explanation is that sexual reproduction introduces greater genetic variation and so enables some offspring to survive better in changing or novel environments. Another theory is that sexual reproduction provides the best defence against rapidly reproducing infectious species like parasites. Yet another is that a major function of sex is to repair damaged genes in the egg and sperm.

Our preoccupation with sex makes sense, since that is why we are here. Should we be sorry that there is not a more attractive explanation for our sexual lives? Evolution has programmed us to be preoccupied with sex to ensure reproduction. Sexual activity goes back a very long way. The first animals known to have copulated were a species of fish, now extinct, some 375 million years ago. They were live bearers, that is, the mother gave birth to live offspring. Before this time the females of species, for example frogs, put their eggs into the water where they were fertilised by sperm deposited by the males. For copulation, where the males insert sperm into the female, modifications in the form of extensions, like the penis from the pelvic region, had to be evolved. With reproduction being linked to copulation, animals had to evolve attitudes to the opposite sex in order to make it work. They had to co-operate with their partners but compete with others of the same sex.

Fundamental to the concept of evolution by natural selection is the idea that if we look back far enough in the history of any two species, we will find they are descended from a common ancestor. This common line of descent means that all living organisms are related to each other. It may thus be helpful to look at sexual behaviour in our closest relatives, chimpanzees, and other primates. As will be seen, they have quite complex behaviour in relation to sex and this feature is biologically determined. It is thus plausible that humans also have specific biologically determined behaviours and skills analogous to those of our closest primate relatives.

Females are the dominant sex in most primate communities and may determine social evolution. Communal living is particularly helpful to females because a group of females is better at finding and defending the food which enables them and their young to survive than individual animals. Having a group also helps in keeping watch for predators. But chimpanzees, our closest relatives, have a patriarchal society in which males are dominant. Usually there is an 'alpha' male who controls the group and keeps order during any disputes. Males seek domination because it brings them better mates, and they try to enlist the support of other males. But the female influence is also important in choosing the alpha male. An alpha male must be accepted not only by other males but by the females. Female chimpanzees will show their acceptance of the alpha male by presenting their hindquarters. Sometimes a group of dominant females will oust an alpha male whom they do not like, and back another male whom they would prefer to see as leader. Female chimpanzees also have a hierarchy, and sometimes young females may inherit their status from their high-ranking mothers. The more dominant females will get together to boss the weaker females. But all this female behaviour is aimed at gaining better access to resources such as food, and not, as in the case of males, to outright domination of the group.

Clear differences exist between the mating systems of different primates. Polygyny, in which one male mates with more than one female while each female mates with only one male, is thought to be the fundamental mating system of animals and is quite common, even in humans. But chimpanzees have a promiscuous mating system: each female copulates with many males and vice versa. Sexual selection is related to male competition and female choice, as females invest far more in their offspring, while males may contribute nothing more than their sperm, though they can provide protection and food. Female primates use a variety of body movements, facial expressions and vocalisations to indicate a desire for sex. Female sexual skin swellings can be attractive to males, particularly in chimpanzees and baboons. Males and females may use similar facial expressions to initiate sex. In female primates, hormones related to ovulation can affect both sexual behaviour and the genitalia, unlike humans where there is little effect. Copulation occurs with the male mounting from the rear. Females respond with orgasm less frequently than do males. Primates manipulate and stimulate their own sexual organs in an activity related to human masturbation. Male–male mounts are common but copulation is rare, and there is no real evidence for lesbianism.

There is great variability in primate sexual behaviour. Males are more interested in mating and try to impregnate as many females as possible, while females are more selective. The basis of female choice of a male for sex is not understood, but males tend to prefer older, not younger, females as mates. There is thus a big difference between chimpanzee behaviour and that of their human relatives. The reason for this difference may be that whereas chimpanzees have a promiscuous mating system, humans typically form longer-term relationships, which means young females are more desirable as mates than older ones. But there may also be a genetic determinant for human males to prefer young females, since, as we will see, this has an evolutionary advantage with respect to health and child care.

Food is an especially important issue for primates, especially for mothers looking after their young. A nucleus chimpanzee group is composed of males who hunt for meat, while females gather insects and may thus have greater manual dexterity. Primate females seem to be keener tool users; for example, they extract ants from anthills using stems from plants. Female chimpanzees exchange meat for sex with males, and males who shared meat with females can be observed to attract twice as many females. For females, who have difficulty obtaining meat on their own, this provides food without the exertions and risks of injury when they hunt for themselves. Females often co-operate to gather food.

The two human sexes have evolved different character-istics to fit their evolutionary requirements. Both sexes must get great pleasure from sex and must want to have children, though both factors are more significant for the female as it is she that bears the child. Care of young children must thus be genetically programmed into both females and males, especially into the former. Both males and females must also be genetically programmed not to mate with close relatives, as this, known as incest, can have deleterious genetic consequences for their offspring. Sexual differences in the brain must control these behaviours.

Breeding with close relatives brings severe genetic defects. The closer the parents are related, the greater are the chances that the offspring will be damaged. When two parents are brother and sister, or parent and child, the likelihood becomes very high. The reason for this is that a child of unrelated parents who inherits one abnormal gene may be unaffected, since there is likely to be a normal gene on the other chromosome, but if close relatives mate there is a high likelihood of both genes being similarly abnormal.

Even animals avoid incest. Chimpanzee females usually leave their community to join another before mating and males may also emigrate. Chimp mothers do not generally mate with mature sons or brothers, and there is evidence to suggest that young females avoid the sexual advances of older males that could have sired them. All of these behaviours serve to avoid incest with its negative genetic consequences. While chimps are promiscuous, several studies have shown that they can sense those to whom they are related, and thus avoid having sex with them. In humans there is a similar mechanism designed to cause the avoidance of sibling incest.

In the past human males seeking desirable mates found that they had to compete with other males, and they still do. Aggressive behaviour was one way of seeing off rivals, used either to drive them away or even to kill them. Since aggression was profitable in terms of finding mates and passing on genes, it could be that this is why men have a genetically inherited trait which makes them act aggressively towards other men. But showing such aggression towards females would be likely to be counter-productive, because it could cause them to be rejected as potential mates and fathers to offspring.

Females did best if they chose good mates and nurtured their children. Since the reproductive success of males is not easy to assess, men were likely to be selected if they looked like being willing to take greater risks than females, claims Daniel Nettle. Women, by contrast, are more anxious than men, which relates to their evolution of harm-avoidance behaviour. Such behaviour might be adaptive for women. A female with dependent young or who is pregnant is more vulnerable to unexpected attack than a male. Protecting her offspring, and her own ability to breed are evolutionary priorities. Any conflict within her group can also be a major threat, as it is difficult for her to leave the group and join another.

A basic difference between the sexes in the importance of social skills and behaviour may be part of our evolutionary heritage, particularly for women. Evolution may have selected men to be more aggressive and more risk-taking.

Scheib and her team say it seems likely that men today find younger women attractive because during the course of our evolution males who chose younger women were more successful in breeding than those who were drawn to females that were too young or too old to conceive. These early successes by men have echoes today, because carefully selected partnerships gave rise to better genes, and it was these that were passed down through evolution. So women today may look for successful men who can best support their children. They use their sex appeal to achieve this end. The anthropologist Donald Symons has observed that people everywhere understand sex as 'something females have that males want'.

Sex incurs a heavy cost compared to asexual repro-duction, as two parents are involved in making one child. But the cost of sex in terms of time and energy is almost trivial for the male compared to the female who will bear a child. It is therefore in the male's best interest to spread his genes by mating with as many females as he can. The female, by contrast, has to be very selective as to which man she chooses. She has in mind the children who might be produced, and wants them to be fathered by a male who will be supportive in bringing them up as well as contributing good genes to the union. Thus her criteria are aimed at selecting the best possible male, not just the first who comes along.

So the mating behaviour of our ancestors probably involved the most powerful and dominant males obtaining access to multiple female sexual partners. Men, more than women, can thus be seen to have evolved a greater desire for casual sex. But for women casual sex can have the consequences of an unwanted pregnancy with all its long-term implications. Females in those ancient times were significantly smaller than males and so were likely to submit sexually to the dominant male's advances and may not have generally understood that sex leads to pregnancy. With time the size difference between males and females diminished and relationships between the sexes started to involve staying with a single partner, and perhaps the beginnings of romantic love. A possible fundamental negotiation, already mentioned, was the female exchanging sex for meat and protection provided by the male, and this could have been the origin of marriage among early humans. We have seen that the exchange of meat for sex occurs in chimpanzees. Early human societies are known to have engaged in meat-for-sex behaviour, with the best male hunters probably having the greatest number of sexual partners. Is today's money perhaps the equivalent of meat?

In ancient societies the most reliable suppliers of food were women who collected it rather than the men who hunted for it. This resulted in women making major technical contributions. Women's food-gathering activities gave rise to the domestication of animals as well as agriculture. It was women working with digging-sticks – one of the earliest tools – to procure food from the ground that led to the planting of crops. According to Veer Gordon Childe, an Australian archaeologist who wrote *Man Makes Himself* (1936) and *What Happened in History* (1942), every single food plant of any importance, as well as other plants such as flax and cotton, was discovered by the women in the pre-civilised epoch. Primitive women were also the first potters, using clay to create cooking and storage vessels. It has been suggested that skilled tool-making and tool use in humans thus began as a largely female domain. All the basic cooking techniques – boiling, roasting, baking, steaming – were developed by women. The social anthropologist Robert Stephen noted in *The Mothers: A Study of the Origins of Sentiments and Institutions* in 1927:

> The weaving of bark and grass fibres by primitive woman is often so marvellous that it could not be imitated by man at the present day, even with the resources of machinery. The so-called Panama hats,

the best of which can be crushed and passed through a finger ring, are a familiar example.

Perhaps the least-known activity of primitive women was their work in the construction of homes, whether huts, skin lodges, wigwams or camel-hair tents. So long as hunting was an indispensable full-time occupation for men, it required women to work cleverly at home. The late Oxford evolutionary biologist William D. Hamilton made a clear comment on the two sexes:

> People divide roughly, it seems to me, into two kinds, or rather a continuum is stretched between two extremes. There are people people, and things people.

Men hunting and fighting, while women attend to agriculture and the home, is a typical pattern in primitive societies. (The hunting experience of ancient men, which involved rapid action, and the foraging activity of women, which required long periods of searching, could explain why traditionally women like to take their time browsing in shops while men would rather go in for what they want and rush out with it.) When ancient man had killed his game, he sometimes sent his woman to fetch it, and even if he brought it home himself, he left it to her to deal with it. The discovery of agriculture by women, and their domestication of cattle and other large animals, enabled men to devote much less time to hunting. Spending more time at home, they began to teach themselves the crafts that the women had already mastered, and to make improvements in their tools. The invention of the plough and the use of domesticated animals to pull it was crucial.

Men's freedom from hunting, together with their greater physical strength, made it easy for them to dominate women, who were committed to caring for children.

However, behaviour that seems to go beyond the single aim of improving or encouraging reproductive fitness, such as altruism, is harder to explain. It is difficult to see a genetic link. But research has suggested that any behaviour that is not directly beneficial to an individual but which potentially benefits that individual's relatives can leave its mark on the legacy of the group. This can be clearly seen in the behaviour of today's parents towards their children. The process of natural selection would have favoured this kind of altruistic behaviour in humans because children carry their parents' genes into the next generation. So there is a sound genetic basis for this sort of parental altruism. Similarly there is also a common advantage to living in an altruistic society.

The evolution of human sex required the development of specific body structures like the penis and breasts. How could the evolution of breasts, which gave milk to the newborn, have come about? Even Darwin found it a very difficult problem. He wrote, 'If it could be demonstrated that any complex organ existed, which could not have been formed by numerous, successive, slight, modifications, my theory would absolutely break down.' Small modifications in the lineage of mammalian ancestors gave rise to lactation and the breasts. It is far from easy to understand. An important piece of evidence was provided by the platypus, a monotreme mammal, which has a patch on its breast that produces milk secretions which the infants suck. Mammary glands are thought to have evolved from some mosaic of the different milk-secreting glands in our ancestors. Among our primate cousins, the monkeys and the great apes, the breasts of females enlarge only slightly during pregnancy and lactation, and in the non-pregnant state remain almost completely flat, with merely the nipple projecting.

So why are breasts so important in the sexual life of humans? Why is their size so crucial for human attractiveness and female sexual self-esteem? It must be related to the change in sexual intercourse, from the male mating from the rear to front-to-front contact. Some anthropologists have proposed that the swell of female breasts is equivalent to the swollen fertile buttocks of our female primate ancestors which attracted males. Human female breasts may have become permanently enlarged due to sexual selection, as they potentially resembled the region of the distended buttocks which are employed to signal appeasement as well as sexuality in primates. Over the course of evolution female breasts became permanently enlarged, and thus a permanent sex symbol to signal that the human female was continuously sexually receptive.

But why then do men have breasts and nipples? This is almost certainly due to males being essentially modified females, as we have seen. Females have everything required for reproduction except some means for getting the egg to start developing. The advantage of having two sexes resulted in the evolution of males to provide sperm. The evolution of males involved modification of female development, but vestigial breasts in the male were no disadvantage, and so they have persisted.

Richard Dawkins has speculated that the absence of the penis bone in humans, although present in our nearest related species, the chimpanzee, and other primates, may be due to sexual selection by females looking for an honest advertisement of good health in prospective mates. Human erection relies on a blood-pumping mechanism, and failure to achieve an erection would be a warning to females of possible physical or mental ill health. It has also been speculated that the loss of the penile bone occurred because humans evolved a courtship pattern in which the male tended to accompany a chosen female all the time to ensure paternity of her children, and thus could enjoy frequent matings of short duration. Primates with a penis bone only infrequently encounter females, but engage in longer periods of copulation which the bone makes possible, thereby maximising their chances of fathering the female's offspring.

It seems clear that evolution led to a number of genetically determined sexually different characteristics in humans, both physical and behavioural. These are likely to include greater empathy and less physical risk-taking in females, and more aggression by males. Both sexes developed sexual desire for the opposite sex and the habit of face-to-face copulation. But what evidence for these effects can be found in the male and female brain?

5
Brain

Here's all you have to know about men and
women: women are crazy and men are stupid. And
the main reason women are crazy is because men
are stupid.

George Carlin

It is claimed by a leading American neuroscientist, Larry
Cahill, that clear sex differences exist in every brain
lobe, including cognitive brain regions. In this chapter I
focus on structural differences between male and female
brains, and will examine some functional differences in
later chapters. As we shall see, brain-imaging techniques
report sex-related differences in visual skills, emotional
responses, sexual arousal, facial processing, memory and
language abilities.

The brain is composed of billions of nerve cells –
neurons – and supporting cells, and the working of the
nervous system depends on the formation of neuronal
circuits in which the neurons make even more billions of
connections with each other. Neurons consist of a cell body
containing the nucleus with its chromosomes and genes,
and extensions, which can be several feet in length, that
transmit or receive electrical signals. From the cell body a
long extension, the axon, may contact distant nerve cells to
which it transmits signals. The axon may be insulated by a
fatty coat of myelin along its length to improve its ability
to conduct the nerve impulse. Damage to this coating is a

feature of multiple sclerosis – a disease more common in women than men. The neuron's insulated extensions form the brain's white matter, while the cell bodies form the grey matter. On the cell body are other, highly branched short extensions, dendrites, with which other nerve fibres make contact, so that a single neuron can receive as many as 10,000 separate inputs. Astonishing!

Understanding the differences between the brains of the two sexes is complex and difficult, and is often further complicated because the differences may be due not to embryonic development but to social experience. Moreover it has been argued that apparent sex differences in the relative size of specific regions of the brain may be accounted for by differences in individual brain volume. During development the brain is influenced both by hormones such as testosterone and oestrogen and by other factors in the environment of the developing brain cells. These can affect which genes are turned on and so which proteins are made, and thus control the development of the brain. The sex chromosomes, particularly the Y, also play a role. Complex regulatory networks that involve hormones and gene activity can promote sex differences in nerve cell connectivity and function. Post-mortem analyses have revealed pronounced sex differences in the distribution of hormone receptors in male and female brains.

A number of genes, including some from the sex chromosomes, are expressed differently in the male and female brain, with hormones again playing a key role. These genes can affect neuronal connections in the brain in subtle ways. The SRY gene, which initiates male hormone expression via the testes, is expressed also in the brain, and six genes on the X chromosome are expressed at higher levels in the male brain. SRY and another Y chromosome gene have been found to be expressed in the hypothalamus, which links the nervous system to the hormone system, and in the frontal cortex of adult males, but not females. At mid-gestation significant sex differences in gene expression are found for genes encoded on the Y chromosome. A dramatic advance was made by a team of Swedish researchers, Trabzuni and team, who reported that 1,349 genes are expressed differently in the brains of men compared with women. The function of these genes remains to be determined. Ingalhalikar and his group found that male connections in the brain go from front to back while those of females go from side to side. They suggest male brains are structured to facilitate connectivity between perception and co-ordinated action, whereas female brains are designed to facilitate communication between analytical and intuitive processing modes.

A very different view of the brain is illustrated by two recent books which largely ignore biological influences and focus on social influences. Cordelia Fine, in her book *Delusions of Gender: The Real Science Behind Sex Differences*, shows an unfortunate misunderstanding of embryonic development and cell biology, arguing that genes do not determine our brains but 'constrain' them. That is wrong, as genes do determine the development of the embryo, including the brain with its neural connections that control behaviour. Worse still, Fine does not seem to be aware of biological differences in sexual behaviour, and of how we are a society of cells that determine everything we do. Another book, Rebecca Jordan-Young's *Brainstorm: The Flaws in the Science of Sex Differences*, is similar and claims that 'whatever is written in our genes must be an open-ended story'. She writes that 'The evidence for hormonal sex determination of the human brain better resembles a hodge-podge pile than a solid structure.' She believes that any brain differences between the sexes have probably developed as a result of social experience but she ignores studies showing such differences in behaviour even in very young children. Both books also ignore the key insights into genetic differences revealed by an understanding of evolution. By contrast a book which sensibly takes biological factors into account is Diane Halpern's *Sex Differences in Cognitive Abilities*.

The evidence for genetic differences in males and females is, of course, overwhelming; one has simply to look at the question of sexual attraction. The brains of men and women may be mainly alike, but they show consistent and important differences with implications for each sex. The differences can influence specific behaviours and may contribute to susceptibility to specific diseases. The male brain is about ten per cent larger than the female, even when the larger body size of the male is taken into account. Men have on average twenty-three billion neurons in the dorsal region of the cerebral cortex, which is largely responsible for higher brain function, compared with nineteen billion in women. The neurons themselves, whether from men or women, show different properties in cell culture and differ in their responses to toxic chemicals. Male brains contain more neurons in specific areas, while the brains of women show greater complexity in certain areas. One cannot overemphasise the complexity of the brain: for its size, it is the most complex structure in the universe.

Differences in brain structure have been investigated by a variety of methods, including post-mortem dissection.

Recent neuroimaging studies on living subjects have accumulated substantial evidence supporting the notion that sex is associated with differences in brain connectivity. Functional magnetic resonance imaging (fMRI) harnesses the paramagnetic properties of oxygenated and deoxygenated haemoglobin, providing images of neural activity due to changing blood flow in the brain. It takes repeated scans, usually one a second, which track the movement of blood through the brain. From this movement it is possible find out which sections of the brain are active and responding to outside events and activities, and to do so 'live'. The images created give a visual representation that reflects which brain structures are activated during the performance of various tasks. For example, a subject may be presented with emotional images, have their memory tested or solve a puzzle. Using powerful magnets fMRI can localise changes in brain activity to regions as small as one cubic millimetre, so it is an excellent tool for the investigation of functioning brains. Another MRI technique is diffusion tensor imaging, which can characterise local microstructure based on water diffusion and show connections between different regions.

There have been numerous attempts to find differences in behaviour that correlate with sex differences in the human brain, but the results so far are inconclusive. There are arguments both for and against the influence of sex on the way that men and women's brains function. It is a controversial area, but it is generally thought that there *are* differences and that these will be demonstrated eventually. As we shall see, there are significant differences in behaviour from an early age.

As Melissa Hines has pointed out, experience can change the brain throughout its lifetime, and the development of new connections between neurons continues into adulthood. A brain sex difference may be caused by experience even if it is known to be influenced by early hormone exposure. Some people with early hormone abnormality have been of key significance in studying neural function or structure. Some neural differences in girls and women with congenital adrenal hyperplasia – CAH – have been of great significance in the study of the role of testosterone in human development. The embryos of individuals with CAH are exposed to high testosterone levels before birth, similar to those of healthy male foetuses. Consequently females with CAH are born with partially masculinised external genitalia and show increased male-typical behaviour throughout their lives. Both normal boys and girls with CAH were found to have smaller amygdalas, involved in emotional memory, but no other differences in brain structure were identified. A brain that develops as male in the embryo is far less likely to display female sexual behaviour in adulthood, even if treated with female-typical hormone replacement.

The cerebrum, including the cerebral cortex, is the largest part of the human brain, and is associated with higher brain function such as thought and the control of movement. The human cerebrum comprises left and right hemispheres, and information is exchanged between them mainly by a bundle of nerve fibres, the corpus callosum. Some small sex-related structural differences have been identified. The two hemispheres look mostly symmetrical, yet there are some controversial studies which have shown that each side functions differently and that in this respect too there are male–female differences.

Of particular importance in relation to sex differences

is the amygdala, which processes and stores memories of emotional events and is also involved in immediate emotional responses. When active it gives rise to fear and anxiety. In the adult human brain, the male amygdala is significantly larger than the female, even taking total brain size into account. The right amygdala shows a greater functional connectivity in men than in women, but the left amygdala shows the opposite trend. There is significantly greater activity in the left amygdala of women remembering emotionally arousing pictures. The stria terminalis serves as an output pathway of the amygdala and connects to the hypothalamus. Its central subdivision, the bed nucleus of the stria terminalis (BSTc) is on average twice as large in men as in women. This sexual difference of the BSTc does not develop until adulthood, at about twenty-two years of age.

Part of the hypothalamus, which links the nervous and endocrine systems, is twice as large in men as in women. The hypothalamus synthesises and secretes a number of neurohormones which in turn stimulate or inhibit the secretion of pituitary hormones, including some related to sex. The sexually dimorphic nucleus (SDN) is a cluster of cells in the hypothalamus that is believed to be related to sexual behaviour. The male SDN is about twice as big as the female because it contains larger cells and more of them. The difference is due to exposure to higher levels of testosterone in the male embryo.

Only a few neural sex differences in the brain have been linked to behaviour. A well-established link involves the third interstitial nucleus of the anterior hypothalamus (INAH-3). The structure of this part of the hypothalamus has been found to be two and a half times larger in men than in women and to contain twice as many cells. This region is linked to gender identity, as we shall see when considering transsexuals. INAH-3 is smaller, and therefore more like that of females, in homosexual than in heterosexual men.

Broca's area is a region with functions linked to speech production and may be larger in women. One region which is certainly larger in women is the hippocampus, which plays a key role in both short-term and long-term memory. Left-hemisphere auditory and language-related regions, compared to overall brain size, are larger in women than in men. This may indicate a structural basis for a female superiority in language processing. Primary visual and visuospatial association areas of the parietal lobes, involved in spatial sense and navigation, are proportionally bigger in men, in line with the reports of superior male visuospatial processing.

The ratio between the size of the orbitofrontal cortex, a region involved in regulating emotions, and that of the amygdala, involved in emotional reactions, has been found to be significantly larger in women than men. This may relate to behavioural evidence for sex differences in emotions, since women are generally considered to be better at expressing their feelings than men, and to have superior skills in bonding and connecting to others. The straight gyrus, a ridge on the cerebral cortex, is about ten per cent larger in women compared to men after correcting for males' larger brain size. It is claimed that a large gyrus correlates with a more 'feminine' personality in adults, irrespective of biological sex. Unexpectedly, however, the gyrus is actually larger in boys than girls.

A distinction is drawn between grey and white

matter in the brain. Grey areas are those which contain more cell bodies, while white matter is mainly made up of axons, the long extensions whose colour arises from the whiteness of their insulating myelin. In the past, studies of white matter used invasive techniques such as dissection and histological staining, and so could only be done post mortem. Modern neuroimaging techniques are non-invasive, and several MRI techniques can be used to investigate white and grey matter. Males have more grey matter in their brain than women, but there are certain regions in the brain where women have more grey matter than men. Grey matter represents information processing centres in the brain, and white matter represents the connections between such centres. The significance of these differences is not fully understood, and the different sizes of the brain complicates matters, but the differences may help to explain why some men are better at tasks requiring local processing like maths, while women are better at integrating and assimilating information, such as that required for language.

Yet although these structural differences in the brain may be related to certain skills, they have no effect on intelligence. Since women have proportionately less white matter but more grey matter, it seems possible that they might make more efficient use of the available white matter. Diffusion magnetic resonance imaging has in fact shown more efficient organisation in the female brain, finding that women have higher local functional connectivity – the degree to which activity in each part of the brain is correlated to activity in every other part. According to Tomasi and Volkow, men have lower brain connectivity compared to women and this might affect functions that require specialised processing, such as spatial orienting, whereas women's higher connectivity may be related to optimising language skills. Indeed, they say, brain imaging studies show that women have higher brain activation and better performance during difficult verbal tasks than men. Women access both sides of their brain more efficiently and can therefore make more use of the right side of the brain. This may explain why they can focus with greater facility on more than one task at a time. They are reported to often preferring to solve problems through multiple activities, and this may be related to claims that they are better at multitasking.

There is evidence that activities which require high levels of training or which make intensive cognitive and motor demands, such as preparing for exams or even juggling, can lead to changes in the amount of grey matter in particular regions. Learning involves changes in the connections between neurons and two things may happen: new connections may form and the structure of existing junctions may change. For example, London taxi drivers are reported to have larger hippocampuses than London bus drivers, because the hippocampus is specialised for acquiring and using complex spatial information such as that needed to navigate efficiently. Taxi drivers have to remember hundreds of routes whereas bus drivers follow a fixed route. Small changes due to genes could also have significant effects on grey and white matter.

Having described some of the differences in brain structure between the sexes, in later chapters I will examine how these might be related to differences in behaviour.

6

Children

Women are nothing but machines for producing children.

Napoleon Bonaparte

Girls and boys are different from birth, not only physically but also psychologically. Clear sex differences with a biological basis can be seen in the behaviour of children from a very early age, implying that these differences were programmed during development and have a genetic basis. A few hours after birth girls are more sensitive than boys to touch, and some forty hours after birth girls look longer at a face than boys, and boys look longer at a suspended mechanical mobile.

Brain maturation in male infants is delayed, and at two to four weeks old they are more easily aroused from quiet sleep, possibly because of this. That female infants should appear to sleep more soundly is consistent with reports of increased sleep disruption and problematic crying in male infants throughout the first six months. The development of sleeping rhythms is slower in boys than in girls. Their motor systems also develop differently. When newborn, boys are less able to control their movements, though later they take part in more active and co-ordinated play.

Girls as young as twelve months make more eye contact than boys of the same age and respond more empathetically to the distress of others, with sad looks and sympathetic vocalisations. Foetal testosterone levels

55

are related to frequency of eye contact at the age of twelve months: the lower the concentration, the more eye contact. This is a nice example of how testosterone can affect brain function. The male preference for focusing on systems, as will be discussed later, is evident very early. At one year research showed boys to have preferred to watch a film of cars than a film of a face expressing strong emotion. Little girls showed the opposite. Female babies are more sensitive to smells, to sweet tastes and to touch, and they engage more in making sounds.

Boys have on average a stronger grasp reflex at birth, and they have greater leg strength from three months and greater arm strength from nine months old. Little boys are more physical than little girls, and two-year-old boys are better at throwing. Six- to nine-month-old boys are more likely than girls to imitate rapid movements such as hitting a balloon. There is thus an early male advantage in some motor skills.

It has been found that when babies are frightened in a strange room at four months old, twice as many girls as boys cry, and at twelve months old in the same situation girls move toward their mothers, while boys look around. The young of monkeys or baboons behave in a similar way which supports the possibility that these psychological differences between girls and boys are the product of biological differences. This makes sense as females are much less physically able to resist attacks of any sort.

Conversations can have a strong positive effect on infants' development, and Cheslack-Postava and Jordan-Young found mothers may offer more verbal stimulation to girls than to boys. Mothers were found in one study to make more interpretations and have more conversation with their daughters, whereas they made more comments and gave more instructions to their sons. This may be because girls have a wider social environment and more opportunities for contact with others, not just with their mothers, either through play, speech or merely eye contact. These interactions are important in helping to develop empathy. Girls begin to gesture when they speak a month or two earlier than boys, at around two years, and social factors like mothers speaking more to girls may be the cause of this.

Sex differences in children's play provide important information. Preference for gendered toys may be first seen as early as ten months, and is definitely present by the age of three, and increases throughout the preschool years. Babies and toddlers spend most of their waking life playing, and there are large sex differences to be seen in their play, such as a preference of girls for dolls and soft toys, and of boys for trucks and mechanical toys. When tested at twelve, eighteen or twenty-four months old, girls have been found to look at dolls much more than boys do, while boys looked at cars much more than do girls. However, in one study made on twelve-month-old infants, both boys and girls preferred dolls to cars, suggesting that older boys' lack of enthusiasm for dolls might develop later, perhaps due to social influences – or perhaps because the study used a picture of a car rather than one that moved. At six years old ninety-nine per cent of girls play with dolls but just seventeen per cent of boys. Boys, instead, prefer mechanically based play at that age.

Contrary to some social theories, there is evidence that these toy preferences and other sex-typical types of children's play are biologically determined and are influenced prenatally by testosterone. Foetal testosterone, when measured in amniotic fluid, relates directly to male-typical play in both boys and girls and predicts sexually differentiated childhood behaviour. Girls with CAH, who were exposed to higher concentrations of testosterone in the womb, have more male-typical play, though parenting and other social factors may play a key role. They play more with what are regarded as boys' toys and less with girls' toys compared to sisters who are unaffected or unrelated controls, and when offered a toy to keep they are more likely to pick a typically male toy such as an aeroplane. Their choice of playmates tends to include more boys than unaffected girls, and they prefer a boy-typical play style. Conversely, in a systematic assessment of boys with low prenatal testosterone exposure, male-typical play was found to be reduced in proportion to the lowering of the hormone concentration. We thus see a biological basis for early differences in the behaviour of our two sexes.

Children prefer to play with their own sex, and this, along with other sex differences, becomes more noticeable from two and a half to five years. Once children start attending school, taking part in play with their peers of the same sex is an important factor. Girls' speech tends to be more co-operative and their conversations can continue longer. Boys meanwhile play more group games than girls. There are differences between girls' and boys' aggression when young. Boys use more force to defend their territory and possessions as they grow older, whereas girls are initially aggressive but become less so and are better able to control their behaviour. This may be as a result of social conditioning.

As Zakriski and her team have pointed out, there are robust differences between older girls' and boys' social environments. Compared with boys in their age groups, older girls more often experience peer talk and adult praise, but they encounter less teasing from adults as well as fewer warnings and punishments. However, girls are observed to be more likely than boys to respond aggressively to direct adult control, warning or punishment, whereas boys are more likely to be aggressive in response to adult instruction. Greater male resistance to instruction has also been noted in toddlers, suggesting it is a key feature of boys' interactions with adults. Girls' reactivity to discipline imposed by mainly female counsellors is consistent with evidence that girls' aggression is pronounced with female caregivers, and suggests that they may perceive such interventions as violations of normal female relationships.

Both boys and girls start to develop more skilful forms of play during early childhood, but sex differences in social play may be influenced by sex differences in mental skills, because girls develop language and the ability to understand other people's minds earlier than boys. Boys will be slower in general to speak their first multiword sentences than girls. Girls are the first to involve themselves in social and structured types of play. By the age of three to four years old girls will get together to play, and by four to five they are able to co-operate with each other. By five to six, they have quite sophisticated social interaction with their peers. By contrast, younger preschool boys are more likely to play on their own than preschool girls. Later, boys play in larger groups often organised in a hierarchy, while girls share more and play in smaller groups.

Moving on to the whole question of gender, parents usually have no doubt as to whether their newborn infants are girls or boys, but abnormalities in development can occasionally make this uncertain. Parents determine a child's earliest experience as to what it means to be male or female since parents treat sons and daughters differently even when they are babies This extends to dressing them in different colours – blue for a boy, pink for a girl – and giving them quite different toys: dolls for girls, toy trains for boys. Almost as soon as babies are born parents have different expectations about the behaviour of sons and daughters. These relationships can have long-term effects on how children behave and augment the differences between the sexes – but biological factors clearly play a fundamental role.

The development of gender identity is quite complex, and while it has a strong biological cause, social factors can play a role when the child is young. The differences in the body and brain caused by hormones in the embryo are thought to underlie sex differences in a wide spectrum of behaviours, including gender role. Gender is one of the first social concepts to be learned, and all sorts of stereotypes can be picked by children from which they can make assumptions about their roles as boys and girls.

From a developmental perspective, when infant boys discover their genitals at eight to ten months of age their awareness of their biological sex can begin. Between one and two years old boys and girls become aware of physical differences between the sexes, and before their third birthday they are easily able to label themselves as either a boy or a girl. At four children have a firm idea of their gender identity and identify themselves as a boy or a girl for life, except in cases of some sexual abnormality. Girls are more advanced in grasping gender identity and half of eighteen-month-old girls understand gender labels like 'lady' and 'man', but most boys at this age do not. Infants begin to use gender labels at eighteen to twenty-one months and most children produce the words 'girl', 'boy' and 'man' by twenty-two months. While there is evidence that social environment after birth has an effect on gender identity or sexual orientation, a major role is played by specific neuronal circuits involving the hypothalamus, such as the third interstitial nucleus of the anterior hypothalamus (INAH-3), which is larger in males and is clearly related to the brain network for male gender identity. Recognition of one's own face appears to emerge at around eighteen months of age and at two to three years old children are able to label themselves and others according to gender without difficulty. These advances seem to be biologically determined as there is no evidence that they are modified by external or social events provided that there are no sexual abnormalities.

The attainment of gender identity requires children to understand that everyone, including themselves, is either a male or a female as Zosuls and her team point out in their study. Measures assessing gender identity have tested children's ability to understand and correctly use gender labels to specify themselves and others. Gender labelling tells us that children have an awareness of separate gender categories and are able to use this information. In order to understand the development of gender labelling in very young children it has been necessary to use largely non-verbal measures. For example, pictures of males and females have been used, sometimes including pictures of the children themselves, and the child has to identify the sex by pointing at a picture of a male or female, or by sorting the pictures, into separate boxes for males or females. Most children are unsuccessful at visually assigning gender labels to categorise themselves or others until they are about twenty-eight or thirty months old. Since girls are more advanced in early language skills than boys, they are able to understand gender categories earlier than boys.

Around the age of two boys and girls begin to play with different gender-typed toys, with girls preferring to play with dolls and boys preferring to play with vehicles. The self-socialisation hypothesis suggests that this kind of behaviour is influenced by boys' and girls' ability to categorise themselves and others by gender. Girls produce gender labels consistently earlier than boys, even when their earlier development in language is allowed for, and greater socialisation with women might explain why they are more successful in recognising differences in gender.

Girls typically have more distinctive clothing than boys, which may make their gender more obvious. Boys' clothing, on the other hand, tends to be more varied. However, all babies, whether boys or girls, are perceived first and foremost as being cute, cuddly and dependent, and little differentiation is made between male and female. But once boys have become physically stronger, more active and independent, parents begin to recognise traditionally male attributes, such as a liking for sports and rough-and-tumble activities, and tend to make gender more obvious to boys with remarks such as 'Don't throw that ball like a girl!'

Traditional developmental theories propose that children acquire gender-typical behaviour through social interaction, which leads to gender identification and to emulating the behaviour associated with their own gender. Males enjoy a socially dominant position in almost all societies and, from an early age, understand that they must uphold that superior position. Girls are taught to cultivate a submissive role. In the majority of societies boys spend more time in recreation, while girls spend more time helping with household chores and child care. In further support of these theories, it is emphasised that boys and girls receive responses of approval or disapproval from their parents and carers when they play with gendered toys, such as dolls, so that they tend to choose objects that have been labelled for their own sex or which they have seen others of their own sex choose. But boys do engage in less male-typed play when with their mothers.

As the work of Dedovic and her colleagues pointed out, early gender socialisation and social learning is an essential factor in determining the different responses to stress observed in men and women, and in differentially emphasising what is important to boys' and girls' sense of self and self-worth. Historically, in Western cultures, boys have been encouraged to be more active and independent, while girls have received closer adult supervision and been encouraged to engage in more dependent behaviour and nurturing play. Higher levels of control from parents may also foster a sense of self in girls that is more socially determined. Gender socialisation and social learning may contribute to girls often valuing social goals above non-social goals and tending to be more interdependent, while boys in general are more likely to be independent in nature. Further, in adolescence girls are more likely than boys to desire closeness, to place a high value on interdependence and caring for others, and to worry about hurting others. Boys, on the other hand, have tended to adopt goals that promote self-interest and autonomy.

Girls with CAH are typically born with varying degrees of genital masculinisation such as an enlarged clitoris and fused labia. Social theorists hold that their abnormal genital appearance at birth could cause parents to treat them differently, and it could be this difference in parental treatment, rather than the neural influences, which alters their gender-typed behaviour. In addition having masculinised genitalia could reduce their self-identification as female, which could in turn cause increased male-typical behaviour. Teenage girls with CAH show less interest in make-up and fashion but more interest in subjects like electronics, cars and sports.

Evidence for a biological basis for male and female behaviour in children is clearly shown from the example of mothers who for medical reasons were treated with hormones during pregnancy. Those treated with progestins which *increased* male hormones had children who showed increased male-typical or decreased female-typical behaviour, while the children of those exposed to progestins which *lowered* male hormones showed the opposite effects.

All this has nice evolutionary support from studies of primates which have found sex-typical toy preferences similar to those seen in children. Male vervet monkeys prefer to spend more time than females handling toys such as cars, and less time with toys such as dolls. Similarly, male rhesus monkeys prefer wheeled toys rather than softer toys. This type of play has been linked to systemising, which is a male-typical characteristic, as I shall discuss in my chapter on emotions. Chimpanzee males in the wild were found to be more likely to use sticks as weapons at all ages, while females of all ages were more likely to treat sticks as dolls – though there were some exceptions. Melissa Hines has reported many studies in non-human mammals where hormones were manipulated during early development to assess the impact of these manipulations on brain structure and behaviour later in life. For example, the female offspring of rhesus monkeys treated with testosterone during pregnancy show increased male-typical, rough-and-tumble play as juveniles, and increased male-typical and reduced female-typical sexual behaviour as adults. These hormone treatments also influence neural structure, enlarging brain regions that are larger in males and reducing those that are larger in females.

Differences in styles of parenting may increase the development of differing interests and skills in boys and girls. In the United States, for example, fathers spend only about one third of the time with their children that mothers do, and so may have less influence. There is evidence showing the effects of parents' attitudes on gender and the modelling of gender roles by siblings. But such gender roles and activities are also strongly influenced by hormones and genes, not only during prenatal development but also later during puberty. Testosterone is the prenatal masculinising agent, but oestrogens probably do not have any effect because both sexes are exposed to oestrogens from the mother. As I have mentioned, it is common for parents to dress their baby girls in pink and their baby boys in blue, although older children regardless of gender prefer the colour blue to other colours. But about a hundred years ago the colours chosen were reversed and baby boys were commonly dressed in pink. Just why there was this change is a cultural puzzle. There is some evidence that although blue is the preferred colour of both sexes, women actually have a evolutionary preference for the colour pink, because recognition of this shade helped them to gather ripe berries and fruit more efficiently. Certainly this preference has been clearly shown in a series of studies of children aged from seven months to five years old, where girls increasingly chose pink, and boys increasingly avoided it. Gender differences also can be found in children's drawings. Flowers, butterflies and women in bright clothing tend to be drawn by girls aged five to six years old, but boys tend to draw more mechanical subjects like cars or trains or soldiers and fighting. Girls who have CAH are more likely to draw more masculine subjects such as moving objects, and to use darker colours rather than feminine brightness.

Once children have learned to speak, boys quite often tell aggressive stories, while girls tell such stories less often. Children under four were given a task to work together for a reward, and boys used physical tactics much more often than girls. One could relate these differences to early differences in empathy. While studies strongly suggest that biology plays an important role, as children grow older, cultural and social factors also play a role in determining a 'male' brain with stronger interest in systems or a 'female' brain with stronger empathy. Social pressures may lead more boys than girls to play team sports and video games and use their hands on practising things like carpentry and car repair, says psychologist Richard A. Lippa. These activities help boys develop specific technical skills. But do these boys follow some sort of biological programming, or are they conforming to social expectations about the role assigned to their sex, as some social psychologists suggest? Gender stereotypes may lead to self-fulfilling predictions, with boys and girls conforming to expectations about the specific abilities attributed to them. Stereotypes about sex differences in abilities can be damaging in certain circumstances, such as the belief that girls are assumed to have lower abilities in relation to maths. But the case for biological influences is much stronger if sex differences are consistent across cultures and if they are unrelated to an individual society's gender roles.

An important and well-established finding is that sex differences tend to be larger in nations that are more gender-egalitarian and economically developed, which suggests a biological basis as there will be much less social pressure for the sexes in those societies to behave in particular ways.

Contrary to the evidence that gender identity has a biological basis, there are people who say that there is no way to make a clear-cut distinction between biological sex and social gender. Some claim that the sexual differences in brain structures and functions do not have a biological basis, but can be due to social experience. There is good evidence that experience alters the development of the brain during childhood and adolescence, and this nerve cell plasticity could in principle help to illuminate the close relationship of sex and gender in an individual brain.

But there is much better evidence for biological differences between the sexes, particularly in the brain, which are specified during embryonic development, as we have seen. By the age of three gender identity is usually firmly established and is extremely difficult, if at all possible, to change after that. A controversial case began in 1967, when an accident during circumcision left one of two twin brothers without a penis. His parents were convinced by Dr John Money, who believed that gender was learned rather than innate, to raise the boy as a girl. The child had cosmetic reconstruction surgery on his genitalia and his testicles were removed. He was treated like a girl and was raised with all the trappings of femininity. But when the parents told their child about the accident when he was a teenager, he then rebelled, started receiving testosterone injections and underwent a surgical reconstruction to become a male again. As an adult he continued living as a man, got married and adopted children, but later committed suicide.

Gender identity is most likely to be influenced by hormones at puberty, according to Berenbaum and Beltz, in a way that is consistent with other links between testosterone and male identity, and oestrogens and female identity. At puberty, which begins around the age of twelve for boys and eleven for girls, there are dramatic sex differences in the expression of male and female hormones, and it is logical to presume that both types of hormones act on the brain at that time. Testosterone level in boys is normally low before puberty and then increases, causing dramatic physical changes. Boys are able to produce sperm, they develop bigger muscles and body hair, and their voices break. During puberty in girls, oestrogen stimulates breast development and maturation of the vagina, uterus and Fallopian tubes. The onset of sexual maturity comes from these changes at puberty, and this leads to changes in adolescents' self-perception and accentuates gender-typed activities and interests as they adopt more adult roles. Precocious puberty, which can occur as early as five years of age, is more common in girls than in boys, whereas delayed puberty is more common in boys. Insensitivity to male hormones or sex chromosome abnormalities in women such as Turner syndrome may cause puberty to fail completely. Girls undergoing puberty show an increase in the total output of cortisol, which is related to stress, while boys show little increase. Increased production of cortisol is normally caused by alarm reactions to stress, and it is possible that these increases may unbalance the stress response in girls at high risk for depression. Depression rates in girls double as they grow up, and this may be one reason why.

The direct action of testosterone on the developing brain in boys and the lack of it in girls are crucial factors in the development of male and female gender identity and sexual orientation. Berenbaum and Beltz have reported that a male embryo develops into a girl with a large clitoris if it has a deficiency which prevents testosterone from being transformed into an active hormone. These children are generally raised as girls. However, at puberty, when testosterone production increases because of their XY genes, this clitoris grows to penis size, the testicles descend and the child's build begins to look masculine and become muscular. Despite the fact that these children are initially raised as girls, the majority change into heterosexual males, apparently due to the organising effect of testosterone on early brain development in the womb. This confirms the greater importance of hormones and genes over social rearing in the determination of gender identity.

Boys who are born with a partly or wholly absent penis are often treated as girls immediately after birth, but in adulthood only about half of these children continue to live as girls. Male gender identity is not dependent on testosterone at puberty. Some individuals with male-typical chromosomes and prenatal male hormone exposure who were castrated in early life and brought up as girls still have a male gender identity. Interesting data on the effects of male hormones comes from XY individuals with 5-alpha-reductase deficiency, which leads to low levels of testosterone in the embryo. Sufferers are genetically male and have male gonads but exterior female or ambiguous genitalia. Berenbaum and Beltz reported that most individuals with this disorder changed gender from female to male at puberty, and that those who changed gender were not always those with the most masculinised genitalia. This is a strong indication of testosterone acting on them at puberty rather than in prenatal development.

Another abnormality is complete male hormone in-sensitivity. Inability to respond to male hormones affects XY individuals who have normal testes that produce tes-tosterone but lack functional receptors in the tissues which would normally respond to it. Because their cells cannot be affected by male hormones they are born with female-like external genitalia, and are reared as girls and remain as such.

Gender can be complicated from a physical perspective, and when it comes to sport there is a problem of males being identified as females and so having a physical advantage. There are many tests that a female athlete who is deemed non-female for the purposes of competition can undergo, but the outcome may still be unclear. Probably the most reliable and important is genetic. The key test relates to the sex chromosomes and a female with Y chromosomes presents a difficult problem as it is necessary to determine if they have been active and so imposed male characteristics. But these tests will not necessarily reveal the person's gender identity.

At adolescence there is an overall gender intensification, and sex differences increase as children grow up; this is rather significant at adolescence and even a bit later, as they enter their twenties. There are reports that girls have more social anxiety and self-esteem problems with regard to social relationships than boys, who are more concerned about achievement. Girls are more motivated towards intimate relationships than boys, and less prone to physical aggression. Children have sexual feelings at a young age. Small boys often get erections after the age of about seven, and at this age begin to touch and caress their sexual organs regularly. By the time they reach puberty, more than half of all males have tried to masturbate, and eighty per cent of all boys have masturbated regularly by the time they are sixteen. Only forty per cent of sixteen-year-old girls, though, have masturbated according to published studies. It is only when girls reach puberty that they may begin to masturbate, but there can be a delay until much later in their teens or even twenties. This may increase to eighty per cent by the age of twenty, four years later than in boys. The evidence on sex differences in adults is, however, much more wide-ranging, and it is to this that I turn in the next chapter.

7

Sex

Why are women . . . so much more interesting to
men than men are to women?

Virginia Woolf

Childbearing is fundamental to the survival and evolution
of humanity, and so in all cultures sex and sexual activity
are found to dominate people's lives. Sexual activity
includes emotions and behaviour pertaining to sexual
intercourse and the production of offspring. It leads to
key differences between men and women, determined by
biological as well as social factors. Individuals of either
sex possess different combinations of these attributes,
which are characterised as 'male' or 'female', and brains
are neither purely male nor female, though usually
strongly biased one way or the other. To fully appreciate
this dichotomy one needs to further consider the nature
of gender and certain divergences from the norm, such as
homosexuality and transsexuality.

An early question when considering human reproduction is whether the desire for children is genetically
determined, or whether it is simply the outcome of
wanting sex. The answer seems to be both, although there
are whole societies among whom it is not realised that
sexual intercourse gives rise to children. Some Australian
tribes, for example, are wholly ignorant of the mechanism
that causes childbirth and believe that the child is
produced by the spirit of an ancestor entering the mother.

They do not make any link between childbirth and sexual intercourse or understand that sexual intercourse is necessary to produce a child. There is a concomitant belief in reincarnation, and it is prevalent over the whole of the central and northern parts of the Australian continent. This is one remarkable illustration of the emotionally based genetic programme in our brains, both male and female, that drives us to have sex. Much sexual activity is not related to producing offspring, but results from a sexual drive that leads to a most enjoyable activity. It also acts in cementing the bond between a couple. The basic drive is genetic and driven by the evolutionary need to continue the existence of the species. However, many modern women in the United Kingdom do not want to have children for a variety of reasons, such as preferring to concentrate on a career. Over half of professional women there aged forty to forty-five are childless: an example of how culture can alter biology.

A prominent difference between the sexes is seen in their respective targets in sexual arousal as the overwhelming majority of men are attracted by women, and vice versa. Men in general are more easily sexually aroused, while women prefer sex to be linked to an emotional attachment. Men are much more sexually motivated than women; they visit pornographic websites more often, although it should be taken into account that the majority of those sites are made for men by men. Adolescent males may initiate sex without emotional involvement, whereas girls may require to be seduced – and it is possible that this divergence has a genetic basis.

It is the desire to have sex rather than the desire to have children that historically has driven reproduction. Does this desire differ between men and women? A *New Yorker* cartoon by Sam Gross nicely illustrates the issue, showing Adam and Eve walking naked in the Garden of Eden, with Adam saying to Eve, 'Hey! I just figured out how we can have a child without using another rib.'

Evolution has made sure that sexual activity is very common. Ninety per cent of men and eighty-six per cent of women have had sex in the past year, and a recent article in the *Independent* newspaper reported that some men said they had had sex with more than a thousand women – or even many more. Wilt Chamberlain, an American basketball player, claimed in his 1991 autobiography, *A View from Above*, to have slept with 20,000 different women, none of them married and none ever impregnated by him, from the time he was fifteen until he wrote the book at the age of seventy. It is, of course, impossible to verify these claims!

There are basic patterns of behaviour for males and females that may have evolved with a genetic basis to ensure reproduction. The woman may feel good when a man indicates that he finds her sexually attractive, and so become willing to engage in sexual activity. Her vagina will become lubricated so that when the man inserts his erect penis into it she will be able to experience this without discomfort. Women have a reflex vaginal response to sexual stimuli, says John Bancroft, former Director of the Kinsey Institute, regardless of whether they find the stimuli consciously appealing. The clitoris, which develops from the same tissue as the male penis, is very sensitive and becomes erect with sexual stimulation – it is the primary cause of female sexual pleasure. For the man in states of sexual arousal the penis becomes erect and erotically sensitive. Insertion of an erect penis into a well-lubricated vagina is an optimally effective means of stimulation. During orgasm his sperm will be ejaculated into her vagina, and this is a strongly pleasurable experience. All these phenomena are primarily genetic in origin.

There are differences in sexual responses and sexual desires between men and women, particularly with respect to orgasm. Though approximately ninety per cent of women report orgasm from some form of sexual stimulation, most women do not consistently experience orgasm from sexual intercourse. About sixty per cent of women report that they usually or always have an orgasm when masturbating compared with twenty-nine per cent who do so during partnered sex. The corresponding figures for men are eighty and seventy-five per cent. Thus most men routinely experience orgasm from sexual intercourse. But whereas men are more likely to orgasm when sex includes vaginal intercourse, twice as many women are more likely to experience it when they engage in a variety of activities such as oral stimulation.

During orgasm in both men and women the hormone oxytocin is released into the bloodstream. For women this leads to a desire to lie still for a while afterwards, thus increasing the likelihood of conception. The typical male orgasm lasts no more than a few seconds, while in women climaxes of up to a minute are known. Males have a higher interest in orgasm, perhaps because they place less importance on permanent relationships.

There is a traditional view that women fall in love first and discover sexual desire later, while men long for sex and only subsequently learn to love. In line with this women quite often say that their most satisfying sexual experiences are based on emotional connection to someone, not just on sexual activity or orgasm. In fact women's ability to feel sexual desire and achieve orgasm may not be a universal characteristic. There are societies, according to Margaret Mead, where women apparently neither masturbate nor feel any particular pleasure through sexual activity, and where female orgasm is so much an unknown phenomenon that it does not even have a name. In these communities it is men who ensure that reproduction takes place, for all they desire is to have sex.

But the function of the female orgasm compared to the male is far from clear. One possibility is the advantage it provides by causing the uterus to suck in sperm. Another explanation is that the pleasure it gives increases women's desire to have sex and also helps bonding with their partner – though as we have seen many women's orgasms are from masturbation, not sex. It is also possible that the female orgasm evolved in a manner similar to nipples in men, and this resulted in some aspects of the male orgasm being introduced into female development. In this respect, the female orgasm could be a male modification.

Men mainly become sexually aroused by erotic images of women. Hard-wired into the male brain, after millions of years of evolution, is a desire for sex in response to the sight of a good-looking young woman. In contrast both male and female erotica cause sexual arousal in women, whether heterosexual or lesbian. They thus have a bisexual arousal pattern, even when they prefer sex with men rather than women. These findings may suggest a fundamental difference between men's and women's brains, and illustrate important differences in sexual emotions in the two sexes.

An important example of genetic programming in females is the menopause, which in British women occurs at an average age of fifty-one. Why do women forgo many years of reproductive life? What selection pressures could result in this adaptation, which is shared only by humans and two species of whale? Primates and other animals have declining breeding rates towards the end of their lives but then die quickly, whereas women continue to live for what may be over a third of their lifespan unable to breed. One explanation is the 'good mother' theory, namely that energy should be devoted to looking after the children a woman already has rather than having more. Then there's the 'grandmother' hypothesis, which contends that women who have stopped ovulating are now freed from the risky costs of reproduction, and so are better able to invest not only in their own children but also in their grandchildren. A recent research project in Gambia by Shanley's team revealed that children are significantly more likely to survive to adulthood if they have help from a grandmother.

Andropause is the name given to a condition in men in their late forties and early fifties that has some similarities with the menopause in women. But there is a steady, not a sudden, drop in testosterone in men from about the age of forty, unlike the sudden drop in female hormones at the female menopause. Some dispute the term and assign other underlying causes such as stress or lifestyle to explain symptoms such as tiredness, depression and loss of energy and concentration. Andropause does not prevent older men reproducing until quite an advanced age, but it may lead to episodes of impotence.

Sexual stimulation using visual images has been used to elicit a state of sexual desire and arousal in men and women alike, and their brains have been found to react differently. Both the amygdala and the hypothalamus showed substantially more activation in men than in women during these studies. In contrast to men women did not show significant bilateral activity in the amygdala and hypothalamus. During tactile genital stimulation of the erect penis, differences in brain activity were particularly marked and the brain regions activated were not the same as those activated by clitoral stimulation. This shows that men and women have different ways of reaching orgasm. But the patterns of activation and deactivation in the brain are largely similar in men and women during the orgasm itself.

On what basis do men and women choose their sexual partners? Is physical attractiveness the essential factor? Humans all around the world discriminate between potential mates on the basis of their attractiveness. Attractiveness has been defined in most societies through-out history by the different clothing worn by men and women, with women expected to wear clothes that are more visually alluring. Visual stimuli play a much greater role in male sexual behaviour. A recent survey of over 10,000 people, in thirty-seven countries on six continents, reported that men everywhere value physical attraction and youth in their potential mates, while women rate ambition, status and financial resources more highly. From an evolutionary perspective, the psychological mechanisms underlying what is regarded as attractive in a mate are adaptations that have developed so as to increase the propagation of healthy offspring.

The main evolutionary hypothesis is that attractiveness reflects an individual's health, and is thus a good indication of their ability to reproduce. Thus the desire to propagate genes beneficial for reproduction would suggest that humans were and are able to distinguish a desirable mate partly by looking at their outer signs of good health. Both obesity and extreme thinness, for example, are generally thought to be indications of poor health in men and women and to be heritable characteristics. Obesity has been known for a long time as a risk factor for a variety of serious illnesses and early death, which is now clinically recognised. There is some evidence that attractiveness correlates with health in women in Western societies, though not in men. Yet no relationship was found in one large study in 1999 by Thornhill and Gangestad, when men's and women's facial attractiveness assessed during their teenage years was compared with their health assessed years later. Although this finding appears to be damaging to the view that perceptions of attractiveness have evolved as assessments of health, the authors did point out that there are problems. Since modern humans live in environments different from those of our ancestors millions of years ago, facial attractiveness might not be associated with health to the same extent as it was. It is clear that modern medicine and healthier lifestyle have both had an effect. In the light of current evidence it is difficult to sustain the idea that facial attractiveness alone can be the full explanation of how we choose our mates.

Another possibility is that attractiveness in general was a signal of a good immune system which can cope well with environmental challenges such as parasites. Three-month-old children gaze longer at faces judged to be attractive than at unattractive ones, and this implies that recognition of beauty is not learned but that we may be born with an innate detector. However, the criteria for beauty in women do vary across cultures, as we shall see. Women are much more concerned with their appearance than men.

It is claimed that symmetry has greater effects on men's preferences than on women's, and that symmetrical women are more attractive and also more fertile. Research has shown that these women have levels of female hormones thirty per cent higher than usual, so there may be something in the reproductive theory. Some other attractive features such as clear eyes and smooth skin clearly denote health, as well as non-facial features such as being in an optimum weight range. Age is important for female attractiveness. The years from age twenty-three to twenty-eight are best for reproduction. Men have evolved to be attracted to women with high childbearing potential, and thus the fact that female fertility typically declines after the late twenties is very relevant, making greater age in women a negative feature for males. The physical beauty of women attracts more attention in most societies than do the good looks of the men, Hollywood excepted. Women face more risk in choosing a mate, and therefore prefer higher-status males. One study of thirty-seven cultures found that women chose mates who were older, committed and resourceful. But women have been found to value highly men with slightly feminised faces, as they appear to be more co-operative and honest, and therefore better potential parents.

An expression of happiness in women is most attractive to men, but one of the least attractive factors for women when judging the appearance of men. A series of studies in which a thousand adults rated the sexual attractiveness of people in photographs demonstrating different body language, such as broad smiles for happiness and averted eyes for shame, found that women preferred men who looked a bit unhappy. Maybe this illustrates their mothering instinct. Expressions of pride, by contrast, were the most attractive male attribute for men, though men rated them one of the least attractive in women. But male smiles did suggest to women that the man would be good to children.

There are claims that women maintain direct eye contact while speaking for about twelve seconds, while men maintain eye contact for only about three seconds. It is also claimed that a woman, whether she lives in a jungle or a modern Western city, signals sexual interest with the same sequence of expressions: she smiles, lifts her eyebrows and looks at the man with wide-open eyes. Then she tilts her head down and to the side and looks away. It could be an innate pattern for flirting. The chest thrusts of men may play a similar role. But all these claims may be anecdotal.

Cross-culturally there is high variability regarding the ideal female body and breast size. In the Renaissance the ideal woman was fatter than at any other time in modern history. By today's standards paintings from that era celebrate women whom today we would consider overweight but at that time were the epitome of sexiness. Body Mass Index (BMI), which is defined as a person's weight in kilograms divided by the square of their height in metres, provides a numeric measure of a person's 'fatness' or 'thinness', and may represent a primary factor in determining physical attractiveness. But there are differences between cultures. Swami and Tovée found that men from industrialised societies tend to prefer a lower BMI than those from semi-industrialised societies. By contrast, in most traditional non-Western countries, body fat is prized as a sign that a person is more prosperous and successful and in women as a measure of femininity. Some cultures prefer plumper women, and in Africa women with ample proportions and a generously sized backside are traditionally regarded as attractive.

Extreme examples of socially determined beauty have occurred around the world. These include the Mangbetu people of central Africa who used to wrap the heads of female infants in giraffe hide in order to induce a cone-shaped head; these cranial deformations have been found also in ancient cultures in the Middle East and South America. In other African societies plates are inserted into young women's lips to enlarge them, or the earlobes of both men and women weighed down so that they become elongated. The Padaung people of Burma considered a very long neck to be the ideal of female beauty and girls were fitted with progressively more and more brass neck rings to achieve this look. This custom still survives among Burmese refugees in northern Thailand. The modern fashion for body piercing among young people, although usually less extreme, may echo this sort of tradition.

A peculiar problem is why in many classical paintings women have exposed breasts, and why women in some societies, including modern ones, are willing to expose their breasts in public, particularly on the beach. There is often opposition from males to this exposure and it is frequently prevented by law. By contrast it is rare for either men or women to wish to expose their genitals in public. However, covering one's body out of modesty is not universal. In the far north of America the Inuit are normally warmly clothed against the cold weather, but it is socially acceptable for them to go without clothing in their snow huts. Come to that, why are breasts attractive and genitals ugly in our society? Topless in women is one thing, but bottomless in men quite another.

That humans, particularly women, are somehow programmed for monogamy is one proposed evolutionary theory. But there is little evidence to support this view since in over 150 cultures infidelity has been cited as the most common cause of divorce. In the West a spouse's infidelity is cited as the primary cause of divorce in between twenty-five and fifty per cent of cases; and approximately half of divorced men and women report that their former spouse had engaged in extramarital sex. Twenty to twenty-five per cent of men and ten to fifteen per cent of women in Western societies are known to have engaged in extramarital sex at least once during their marriage, says the Kinsey Institute. Pregnancy is a particularly dangerous time for male infidelity. Women prefer to cite emotional reasons for it, such as 'falling in love', and are less willing to accept purely sexual justifications for extramarital affairs. Women fantasise about sex with another partner in terms of personal qualities, whereas men concentrate on physical characteristics. Men can even have sexual pleasure when committing rape, since their sex drive may allow them to ignore the feelings of their victim.

There is a strong biological basis for avoiding incest, as discussed earlier. What clues are used to make sure that incest does not occur? The Westermarck hypothesis is that we develop a strong sexual aversion to people with whom we have been reared in close proximity in early childhood. A selection pressure among animals exists for fathers to identify potential daughters so as to avoid perceiving them as potential mates. In human societies, as shown for example in the data gathered from a kibbutz in Israel, it is highly unusual to marry an unrelated person with whom you have been brought up, but there is nonetheless evidence that as adults these people may develop a strong sexual mutual attraction. Evolutionary psychologist Debra Lieberman reported after a survey of 600 people that there is some form of brain mechanism which assesses various clues to estimate how closely related two people are. This is manifested more strongly in women because women invest more in the genetic fitness of a particular child than men do. On the other hand, men who grew up with only brothers do not seem to find the idea of incest as inherently distasteful as those who grew up with sisters.

Some experts have calculated that as many as one in a hundred women have been sexually abused as children by a family member, though it is difficult to find precise numbers because shame leads to under-reporting. Father–daughter incest is not uncommon, nor the abuse of younger siblings by older brothers. Sex between parent and child and between siblings is almost universally forbidden but some cultures allow sexual and marital relations between aunts and uncles and between nephews and nieces. There are cases, such as the Egyptian pharaohs, where the elite have accepted brother–sister marriages.

In human courtship there is much emphasis on feminine passivity in contrast with masculine pursuit, and this occurs throughout the animal kingdom. The male is the more active and aggressive in mating behaviour, with the female being pursued and persuaded. This probably has a biological basis, but Margaret Mead's studies among the Tchambuli in Papua New Guinea contradicted this and she found 'a genuine reversal of the sex attitudes of our own culture', the women being 'dominant' and 'impersonal' and the men emotionally dependent'. Mead concluded on the basis of studies of three primitive societies that the 'passivity' hitherto regarded as typically feminine can no longer be regarded as an essential part of the sex. 'The material suggests that we may say that many, if not all, of the personality traits which we have called masculine or feminine are as lightly linked to sex as are the clothing, the manners, and the form of headdress that a society or a given period assigns to either sex.' However, some later researchers have strongly disputed her methods and the robustness of her findings, particularly Derek Freeman in his book *Margaret Mead and Samoa: The Making and Unmaking of an Anthropological Myth.*

It is commonly believed that men think about sex much more often than women. But the evidence is not overwhelming. A study of college students who kept track of their thoughts relating to food, sleep and sex for one week found that the young men did indeed think more about sex than the young women did, but they also thought more about food and sleep. In modern society it has been observed that the average man will spend about forty minutes a day staring at the bodies of ten different women, which is a lot of time. Meanwhile women on average look at six men, and mainly at their eyes, for half that time, just over twenty minutes a day.

Men's sexual fantasies tend to be more explicit than women's, whose imagined scenarios tend to be more complex and personal. Studies report that about half of men think about sex every day or several times a day, which fits with my own experience, while only twenty per cent of women think about sex equally often. The first sexual fantasy occurs generally between eleven and thirteen years old, with men generally starting younger than women, says the Kinsey Institute. Women, when recalling their sexual fantasies, more often see themselves being dominated, while men are the opposite. They see themselves as being in charge, either with a single or multiple partners. The sites of sexual desire in the brain have been identified with fMRI and several of them are in the amygdala.

Departures from the norm in sexual orientation provide some interesting insights into male–female differences. Sexual orientation refers to the gender to which a person is attracted – to the opposite sex if they are heterosexual, to the same sex if homosexual, or to both sexes if bisexual. Estimates of non-heterosexual orientation suggest it occurs in about three to six per cent of men and one to four per cent of women. A representative survey of Britons found that one and a half per cent called themselves gay or lesbian and half a per cent bisexual. About half of lesbians and homosexual men reported the desire to have children.

About twenty per cent of men and women have some erotic feelings for the same sex at some stage in their lives. Lesbian and heterosexual women are physically aroused by sexy videos and pictures of both men and women, while both gay and heterosexual men respond only in accordance with their own normal sexual orientation. Heterosexual women's brain activity is greater when seeing male than female genitals, and vice versa for lesbians.

Gender identity and sexual orientation are programmed into our brains during development of the embryo. There is evidence, though, that the postnatal social environment can have an effect on either, say Bao and Swaab. Sexual orientation is determined during development of the embryo under the influence of genes and sex hormones on the developing brain, and becomes overt during puberty under the further influence of sex hormones, as described in the previous chapter. As development of the genitals takes place earlier in embryonic life than the sexual differentiation of the brain, these two processes can be separately influenced. Sexual orientation is not affected by hormones in adulthood, and removal of an ovary or testis does not influence it, nor does adult treatment with sex hormones. Numerous studies have clearly shown that both gay men and lesbians have a normal concentration of sex hormones typical of their genetic sex. The difficulty of changing a person's sexual orientation, as distinct from gender, is an important argument for it being strongly biologically based.

Homosexual adults are likely to show unusual behaviour by the norms of their gender in childhood, but their home environment and the influence of parents have little impact on their sexual orientation. People's basic sexual orientation is stable and rarely changes except around puberty. Homosexual relationships between young and old are found in primitive societies like New Guinea, where they occur between a young male and an older teacher or tribal holy man. Boys engage in sex with older men for some years, but become heterosexual adults. This homosexual activity is part of the initiation into adulthood of the younger male and it comes to an end when the boy is married. Homosexual behaviour for many people in primitive societies may be merely a kind of sideline, co-existing in people who are predominantly heterosexual as adults.

The direct effect of testosterone on the developing human brain is the main mechanism responsible for male sexual orientation. It is when things go wrong that this becomes clear. Complete insensitivity to testosterone may be caused by mutations in the gene for the hormone receptor. XY males then develop as women and experience straightforward female heterosexual orientation. Despite their Y chromosome their sex-typical behaviour cannot usually be distinguished from that of girls and women in general. It has been suggested by a number of people that there is a relationship between left- or right-handedness and sexual orientation; they report that heterosexual individuals are somewhat more likely to be right-handed than homosexual individuals. But this has not been verified. Neither has sexual orientation been fully linked to variations in prenatal testosterone examined from maternal blood or amniotic fluid. But one characteristic has already been mentioned as providing an indirect measure of prenatal testosterone exposure: the ratio of the length of the index finger to the fourth digit of the hand (2D:4D), which is greater in women than in men. An online study of more than 200,000 people, who measured their own 2D:4D and reported their sexual orientation, found that 2D:4D related as predicted to sexual orientation in men, but not in women.

Women with CAH who were exposed to excess testosterone in the womb tend to have more sexual interest, arousal and fantasy relating to women, and to have less sexual interest in and experience with men, according to Berenbaum and Beltz. If exposed to very high concentrations of testosterone in the womb, they show masculinisation of their genital structures and a greater a variety of male-typical behavioural traits such as aggressive play, but they also display an increased probability of interest or participation in homosexual relationships. And some studies found up to thirty or forty per cent of CAH girls have some form of homosexual attraction. But this still leaves a majority of CAH women who are heterosexual. By contrast the lack of an effect of homosexuality in men of female characteristics, like empathising and not systemising, may be surprising given how strongly these differences are found between heterosexual females and heterosexual males, as we shall see. In relation to certain skills discussed later, some studies have found that heterosexual men outperform homosexual men on tests of typically male skills such as mental rotation and judgement of line orientation, while in contrast, homosexual men show significantly better object location memory (on which women normally perform well) compared to heterosexual men.

There is a strong genetic component to homosexuality. If being homosexual were strictly genetic, then in identical twins there would be a hundred-per-cent correlation rate for sexual orientation. But one study found only a fifty-two-per-cent correlation for male identical twins and twenty-two per cent for male fraternal twins. A study on female twins came up with similar results – if one identical twin was a lesbian, in forty-eight per cent of cases the other twin was also, and for fraternal twins the figure was sixteen per cent. It is, however, still unclear exactly which genes are involved in determining homosexuality. Some genetic studies have suggested an X-linked inheritance, because the X chromosome has genes involved in sex, reproduction and cognition. There are studies suggesting that the gene Xq28 may play an important role in male homosexuality, but it is a controversial field. There is also some evidence that women with homosexual sons have an extreme skewing (non-random inactivation) of one of the X chromosomes, which is normally randomly inactivated. But since the whole process of the development of sexual orientation is so complex, it probably involves many genes.

A surprising finding is that the odds of a boy being gay increase by one-third for each elder brother he has. Fraternal birth order appears as a prenatal cause of fifteen per cent of incidences of homosexuality in males. There are strong social influences determining homosexuality. Gay men and lesbian women have reported significantly higher rates of childhood molestation than heterosexual men and women. A quite different social finding is that many gay men and lesbian women lost their father through death or divorce by the age of ten. Eighteen per cent of gay men and thirty-five per cent of lesbians were found to be affected in this way in one study.

Does the structure of the brain reflect sexual orientation? Differences have been reported in the size of brain regions based on sexual orientation. The third interstitial nucleus of the anterior hypothalamus (INAH-3), which is about twice as large in men as in women, is smaller in gay men and more similar in size to that of females. Heterosexual and homosexual men also have been reported to differ in the size of part of the corpus callosum which connects the two brain hemispheres and allows communication between them. Other research involving brain measurements has also indicated that there may be a neurobiological basis to some sexual orientation in both heterosexual and homosexual men and women.

Transsexuality is different from homosexuality and is

shown when an individual is absolutely certain that he or she belongs to the opposite sex; this may lead him or her to seek sex-reassignment surgery. Transsexuality can be caused by hormones and chromosomal abnormalities. A study of male-to-female transsexuals found that they were more likely than normal men to have a long version of a gene that reduces testosterone binding to cells in the developing embryo. In the brains of male-to-female transsexuals, the nucleus of the stria terminalis (BSTc) was found to be characteristically more female in size and in number of neurons. Similarly the INAH-3 in male-to-female transsexuals was found to be small and of female size and cell number. Conversely a female-to-male transsexual subject was found to have an INAH-3 volume within the male range, even though treatment with testosterone had been stopped three years before. Another female-to-male transsexual subject had a BSTc and INAH-3 with clear male characteristics. This sex reversal of the INAH-3 in transsexual people may indicate an early atypical sexual differentiation of the brain, and changes in both the INAH-3 and BSTc may be part of a complex brain network related to gender identity and sexual orientation. The present data do not support the notion that brains of male-to-female transsexuals are feminised, but they do have smaller BSTc regions than both heterosexual and homosexual men. On the other hand, female to male transsexuals were shown in a study to have four regions of white matter, which normally differ significantly between the sexes, in areas that resembled a male brain, indicating that they were masculinised. A girl with CAH will have a small increased chance of becoming transsexual. About three per cent of women with CAH want to live as men in adulthood, despite having been reared as girls.

There are a few reports of families where there are several transsexual individuals. A handful of twin studies have been conducted, but they are inconclusive and have shown differing rates for transsexualism. But some trans-sexuals show chromosomal abnormalities and all these cases involved the sex chromosomes. A major output pathway of the amygdala and part of the hypothalamus are about twice as large in men as in women, but in male-to-female transsexuals these differences were absent.

Men have the distinction of being responsible for abnormal, criminal sexual behaviours like rape, which might be seen as a kind of perverted sexual strategy to gain access to females, and paedophilia, the causes of which are not yet clearly understood. An estimated 85,000 women a year are victims of rape or sexual assault by penetration in England and Wales according to a 2013 report by the United Kingdom's Ministry of Justice. One in twenty women aged between sixteen and fifty-nine is raped or suffers a serious sexual assault before the age of sixty, and one in five women has suffered some form of sexual molestation. There do not seem to be any strong biological mechanisms for preventing sex between adults and children except with respect to incest. Abnormal sexual behaviour is shown by paedophiles who have distinct characteristics and whose preferred sexual objects are children. What causes paedophilia is not known, but recent investigations have concentrated on the brain and its functions. Male paedophiles tend to have lower IQs, possibly suffered childhood head injuries and have been shown by MRI scans to have differences in their brain structure. But in some societies such sexual activity has been thought acceptable. When Captain Cook visited Tahiti he was astonished to find that the Tahitians 'gratified every appetite and passion before witnesses' by having sexual intercourse in public. He reported in his *Account of a Voyage Around the World* (1769):

> A young man, nearly six feet high, performed the rites of Venus with a little girl about 11 or 12 years of age, before several of our people and a great number of natives, without the least sense of its being indecent or improper, but, as appeared, in perfect conformity to the custom of the place. Among the spectators were several women of superior rank who . . . gave instructions to the girl how to perform her part, which, young as she was, she did not seem much to stand in need of.

8

Emotions

I would rather trust a woman's instinct than a
man's reason.

Stanley Baldwin

It is widely believed that women are more emotional than
men, and this gender stereotype is supported by a body of
scientific research. Women show their emotions more than
men and are more facially expressive for both positive
and negative emotions. They are also better than men in
reading emotional facial expressions both in adults and
children, and they show a stronger emotional response
than men when looking at or listening to infants. Women
also use emotional terms to describe themselves and others
several times more frequently than men. They express more
emotions more frequently and have stronger emotional
responses when viewing images of facial expressions such
as fear, sadness and embarrassment. Men, by contrast, are
believed to feel and express anger and pride more often
and more strongly, and we shall see this particularly in
relation to aggression.

A major difference between the emotions of men and
women lies in this expression of aggression, for which men
enjoy a pronounced physical advantage. Men, as already
mentioned, are more physically aggressive than women,
and men commit about ninety per cent of all murders
and almost all sexual crimes. Crime statistics show that
women account for only about six per cent of prison

inmates, irrespective of nationality, culture, religion and age. One recent survey found that more than one in four women experience violence from an intimate partner at some stage in their lives. The greater frequency of male physical aggression is one of the strongest behavioural sex differences, regardless of many different age groups and cultures. It has an evolutionary origin, as discussed, and men tend to show less aggression against women than against other men. On the other hand, women can be equally aggressive towards men and other women, but their aggression tends to be much less physical as they are more likely to fear all kinds of events that might involve a physical injury. Women can be aggressive by using a number of non-physical means.

Aggression is linked most strongly to the amygdala, which is larger in men, and when influenced by testosterone can trigger aggression and stimulate competitiveness. The prefrontal cortex is the decision-making executive centre of the brain, and it controls emotional information and can put a check on the amygdala. The prefrontal cortex is larger in women and this enables them to look for solutions to conflict more effectively. As Ohrmann and his co-authors point out, significantly, stronger activation of the amygdala was observed in women than in men while looking at angry, fearful, or neutral facial expressions, but not at happy ones. Prenatal testosterone exposure directly relates to physical aggression and many studies have demonstrated this. Girls and women with CAH are more like males, with an increase in physical aggression and increased activity in the amygdala in response to negative facial emotions. The hormone cortisol increases aggressive behaviour in women, but not in men. So there is support for the view that healthy men and women derive their aggression from a different biological basis. There are also significant gender differences in the frequency of physical aggression from as early as seventeen months old, with five per cent of boys but only one per cent of girls often displaying physical aggression.

Boys tend to keep themselves to themselves and adopt wider interpersonal distances, and use their body language to assert dominance differently from females as described by Vigil, and men show dominance through forward-leaning postures, and approaching rapidly. Girls and women show dominance in a completely different way by tilting their heads down, using intense eye contact and smiles in accordance with their partner's responses.

According to Kuschel, on Bellona Island in the Solomons, there is a culture based on male dominance and physical violence, and women tend to get into conflict with other women more frequently than with men. When in conflict with men, instead of using physical means, they make up mocking songs which spread across the island and humiliate their target. If a woman wants to kill a man, she will either convince her male relatives to kill him or hire an assassin. Although these two methods involve physical violence, both are forms of indirect aggression, since the aggressor herself avoids getting directly involved or putting herself in immediate physical danger.

Almost the opposite of aggression is empathy, an emotion which marks a fundamental difference between the two sexes. Empathy is the ability to share others' feelings and to take a positive interest in them, and involves the ability to decode non-verbal emotional cues, as has been well explored by Simon Baron-Cohen. Responding to other people's emotions and sharing emotional responses are fundamental human characteristics. Proverbio and his team pointed out that, because human babies are more vulnerable than other primate infants, knowing how to interpret and respond to crying and other forms of infant communication is crucial, and has an evolutionary origin in women since it is adaptive. Infant survival will depend heavily on a mother's responsiveness to her offspring's visual and auditory signals.

He adds that an example of greater female empathy is their early interest in infants, which has been observed by anthropologists in virtually all human societies, whereas interactions between boys and infants are far less common. The behaviour of non-human primates strongly suggests that this early female interest in infants is a biological adaptation rather than a product of human culture and socialisation, and is supported by comparative evidence; rhesus macaques, for example, show a strong sex difference in interest in infants which persists throughout adult life.

As we have seen, girls as young as twelve months old respond more sympathetically to the distress of other people. But one study has shown that one-year-old boys preferred to watch a film of cars rather than one of people's faces showing strong emotional expressions. By twelve months girls make more eye contact than boys, a difference partly determined by lower prenatal testosterone levels. Making eye contact in this way may be related to sociability and empathising. Female toddlers are more empathetic, showing concern for the distress of others, and heart rates in girls increase when they are told sad stories.

Adolescents aged between thirteen and sixteen years were evaluated by Simon Baron-Cohen and Sally

Wheelwright for empathy, and there was a greater empathic response in girls than in boys of the same age, the difference increasing with age. In this study empathy was measured by a set of forty statements to which the subjects answered how strongly they agreed or disagreed. The sort of statements were:

> I can easily tell if someone wants to enter into a conversation.
> I find it hard to know what to do in a social situation.
> It does not bother me too much if I am late meeting a friend.
> I often find it hard to judge if something is rude or polite.
> When I was a child, I enjoyed cutting up worms to see what would happen.

One may doubt the reliability of such a system of measurement but there is evidence supporting it.

Another key emotion is systemising, a natural ability to analyse data and to organise it into a coherent and logical system. Baron-Cohen's theory is that the female brain is predominantly hard-wired for empathy, while the male brain is predominantly hard-wired for systemising, that is, for understanding and building systems. Empathising and systemising thus differ to a significant degree between the two sexes. So, for example, it has been suggested that a systemiser in our culture will probably choose magazines dealing with computers, technology and science, whereas an empathiser will choose fashion, romance and beauty. In support of these stereotypes it has been noted that psychiatric illnesses which manifest a lack of empathy, such as autism, are far more prevalent among males than among females. Many of the social differences between men and women may be influenced by these two key traits. It is essential to recognise that not all males have 'male brains' or all women 'female brains' with respect to empathy and systemising. There is much variation, and some men have 'female brains' and some women 'male brains'. In general, however, the trend may be summarised as males tending to think narrowly while females think broadly. Boys born with an insensitivity to testosterone are worse at systemising, and girls born with CAH, with their high levels of testosterone during development, have enhanced systemising capability and lower empathy. The higher the levels of testosterone received by an embryo, the greater the chances of reduced empathy. A single administration of testosterone to a female embryo can lead to a significant impairment of empathy in later life.

Behavioural findings and neuroimaging support the idea of general sex-related differences in empathising in the function of the brains of men and women. Women react more strongly than men when looking at painful stimuli and are therefore more empathetic. An experiment by Proverbio involving twenty-four Italian students showed changes in the brain differed between the two sexes when shown the same pictures of humans in positive or negative contexts. The brain response to suffering humans was greater in women and resulted in increased activity in the right amygdala and right frontal area. This was observed only in women and reflects their empathy.

Testosterone affects sociability. In a 2005 study by Knickmeyer and his team carried out in England, boys and girls were compared at four years of age on the quality of their social relationships. This comparison included a popularity scale on which they were judged by how many other children wanted to play with them, and little girls won by miles. The same four-year-old children had their testosterone levels measured when they were developing as an embryo at between twelve and eighteen weeks' gestation, at the time when their brains were being specified as male or female. Those with the lowest testosterone exposure had the best social relationships at four years old. All of these were girls, and this may be related to their capacity for empathy.

A large study of personality differences in some 10,000 subjects found key differences according to fifteen personality scales. Women scored higher on sensitivity and warmth but also on anxiety, while men got higher scores on emotional stability and vigilance, but also on aggression and dominance. The authors of the study, Del Giudice and his team, said: 'Sensitivity differentiates people who are sensitive, aesthetic, sentimental, intuitive, and tender-minded from those who are utilitarian, objective, unsentimental and tough-minded.'

One survey in the United Kingdom found that women are more selfish than men; they are less likely to return a favour or to hand back money after seeing someone drop it, they are more likely to ignore charity workers at the front door than men, and they are more inclined to make negative remarks about their friends behind their backs. But there are also claims that men are more selfish than women.

With respect to telling lies, men are ahead. A survey commissioned by the Science Museum of London in 2010 and carried out by Onepoll asked 3,000 people about their truthfulness. According to the responses, on average, British men tell three lies every day while women, on average, tell only two. Their lies were similar in many ways, but men lied to their partners most often about their drinking habits, while women most often lied to hide their true feelings. Typical male lies were: 'I had no [mobile] signal', 'No, your bottom doesn't look big in that', and 'You've lost weight'. Female lies included: 'Nothing's wrong, I'm fine', 'I don't know where it is, I haven't touched it', 'No, I didn't throw it away', 'I've got a headache', and 'It was in the sale'. Both sexes told several of the same lies, including: 'It wasn't that expensive', 'I didn't have that much to drink', 'I'm on my way' and 'It's just what I've always wanted'.

Schulte-Rüther's team found that there are rather small differences between men and women in their ability to recognise the emotions of other people, but in women the awareness of the feelings of others is accompanied by stronger emotional resonance, while men tend to have a more cognitively driven and distant approach to the emotional states of others. Women show brain reactivity when aspects of a speaker conflict with the content of his or her message – for example, smiling when giving bad news; an effect found much less strongly in men. Women use more facial expressions than men when it comes to showing both positive and negative emotions. The extent to which this has a biological or a social basis is far from clear, but it may be related to empathy.

There is evidence for a biological basis for women's greater sensitivity and vulnerability to adverse and stressful events. The enhanced brain activity in women when viewing suffering humans was recorded in brain regions thought to be part of a mirror neuron system which is believed to support empathy. Mirror neurons fire both when a person carries out a particular action and also when that person sees the same action performed by someone else. There is evidence that overlapping brain activation patterns occur when this happens.

Falling in love involves, to different degrees, sexual attraction and a desire for an intimate relationship. Women are more orientated to friendship-based love, and men to game-playing love, in which they are less dependent and may be more deceptive. Falling in love lowers men's testosterone levels, while it increases women's, and these hormonal changes may result in a temporary reduction of sex differences in behaviour. Becoming a father has also been shown to lower men's testosterone levels. Sexual jealousy is common, and Levy and Kelly reports that studies from around the world have found that men are, in general, more jealous of sexual infidelity than emotional infidelity, while women are more jealous of emotional cheating than sexual cheating. A possible evolutionary cause for this gender difference is that men can never be sure they are a child's father, while women are most concerned with securing a genuinely loyal father to care for children. Women generally report greater disgust than men, especially sexual disgust. While women are more likely to suffer stress due to problems in personal relationships, men more commonly develop stress due to their work. For the most part women tend to use their emotional resources to cope and look to friends for help, whereas men tend to focus on problem-solving and taking action.

Women are more stressed by social exclusion than men, who are stressed by possible future tests of their intelligence. Men tend to adopt a more aggressive response to stressful situations while women prioritise taking care of themselves and their children. A reason for these different reactions to stress is probably based on hormones. The hormone oxytocin, which is best known for its roles in female reproduction and which as we have seen plays a role in orgasms, is released during stress in both men and women. In women, however, oestrogen tends to enhance oxytocin, and the result is that they become calmer and get together with other women for support. The bonds that form help fill emotional gaps and lower risks. By contrast testosterone, produced in high quantities by men under stress, reduces the effects of oxytocin and causes a fight-or-flight response. Cortisol, which is released by the adrenal gland, increases in women exposed to psychosocial stress, and, because cortisol increases appetite, long-term or chronic stress can lead to weight increases.

Across many real-world domains, men engage in more risky behaviour than women. A study by Ginsburg found that boys took more physical risks than girls at both sites when visiting a tube chute and a suspension bridge. Other previous studies had showed that boys were more likely to approach and touch environmental hazards than girls. Sex differences are thus pronounced in attitudes to risk, with men demonstrating more inclination to seek attention and less sensitivity to punishment, and women being consistently more punishment-sensitive. Women experience more daily stress, and often these are family problems and health-related events relating to others, such as children. Men are more frequently affected by stressful events related to work or money.

Goldstein and his co-workers have described how male

and female brains differ in activity in response to stressful stimuli, and in women there is attenuation in cortical arousal circuitry that differs from that found in men. From an evolutionary point of view, it is important for the female during mid-menstrual cycle to judge optimally whether an approaching male represents an opportunity for successful mating or for fight or flight. Thus females may have been endowed with a natural hormonal capacity to regulate the stress response. From an evolutionary point of view this mechanism would have been inappropriate for the male, whose prime responsibility was for protection of the species, necessitating a constant fight-or-flight behavioural response. When under stress, women tend to have increased heart rates, and men tend to have greater changes in blood pressure.

As Kret and her colleagues have pointed out, brain areas involved in processing social signals are activated differently by threatening signals from male and female facial and bodily expressions, and their activation patterns are different in men and women. Male participants pay more attention to the female face, as shown by increased amygdala activity. They show a clear motor-preparation response to threatening male body language, while women do not. Testosterone level in men is a good predictor of aggressive behaviour when looking at angry or fearful male faces. By contrast, as we have seen, oxytocin can cause relaxation and sedation as well as reduced fearfulness and reduced sensitivity to pain and it may be vital in the reduction of the fight-or-flight response in females. There is also a physiological reaction in the response to stress or danger which is different in men and women. If you come up quietly behind a women and say 'boo', she will probably have quite a violent physical reaction as her autonomic nervous system, which controls involuntary body functions, has a lower threshold of arousal to threat. Men are less easily frightened than women.

Women do better than men at recognising bodily action and emotion, as well as facial emotion. They are able to understand subtle facial expressions better and faster than men, though no difference has been found between male and female participants in recognising highly expressive stimuli. Men are less able to identify negative emotions, and are less accurate in distinguishing negative facial expressions such as fear, disgust, sadness or depression. But both sexes are better able to process female facial expressions than those of males, probably because female expression tends to be more exaggerated. Interpersonal communication through non-verbal emotional cues is also more developed in women than in men. Women are quicker and more accurate at identifying changing vocal intonation, and at understand non-verbal communication like small changes in facial expression. This could be genetically determined as females have evolved to care for children, and these characteristics are present in children at a very early age.

Girls and boys under the age of twelve cry with the same frequency, but then a difference develops and women cry more than men. Men and women cry differently. Women are more likely to make crying noises and produce streams of tears down their cheeks, while men cry more quietly and shed fewer tears. The reason for the difference is probably physiological because men's tear glands are smaller than women's; though there is also a social inhibition. Women smile more than men, and Azim and co-workers demonstrated in an imaging study that women activate the parts of the brain involved in language processing and working memory more than men when viewing cartoons. They were also more likely to activate with more intensity the part of the brain that brings pleasure in response to new experiences.

There are gender differences in the tendency to pull funny faces and mimic others, with women doing it more than men. Viewing pictures of people, in either a positive or a negative context, generates a much stronger response in women than in men, as we have seen. Men look at a reflection of themselves in a mirror as often as women do – but women look for longer. So a woman spends an average of two years of her life looking at herself in the mirror; a man spends only six months. As Hamann has made clear, women on average retain stronger and more vivid memories for emotional events than men. Women can recall emotional memories more quickly and can recall more of them in a given period of time, and the ones they recall are richer, more vivid and more intense. In general women tend to experience better enhancement of their memory by emotion. The fact that emotional memories tend to be stronger for women may help to explain why they have a greater prevalence of depression and some types of anxiety disorders. Men tend to recall events using strategies that rely on reconstructing the experience, such as scoring a hole-in-one in golf. They more easily recall important experiences that are associated with competition or physical activities. There appear to be differences in the structure and chemistry of the brain for this kind of memory which are affected by hormones. The hippocampus, which is the area in the brain most involved in memory, reacts differently to changing levels of male and female hormones.

It has been suggested that the things that demonstrate a solid personal relationship are quite different for men and women. Men tend to feel closer through shared activities like sports, whereas women are more likely to feel closer through communication and intimate sharing of experience. Women are very good at making relationships that are meaningful and long-lasting. They work very well towards a desired result when collaborating with others. They will encourage someone to complete an assignment, while men are more apt to show a person how to complete the project, the differences possibly respectively reflecting empathy and systemising.

As Vigil describes, compared to boys and men, girls and women report higher levels of negative life experiences, lower self-esteem, and more symptoms of depression following a traumatic experience. This characteristic may be especially emphasised during adolescence, but average levels of self-control are significantly higher among girls than among boys. Women are significantly more positive toward their own gender than men. This effect is already present in primary-school children. Women are more susceptible to negative emotions in their lives as we have seen. Evidence that women apologise more than men is probably explained by men rating their own offences as less blameworthy than a woman would. Women tend to use complaints as a plea for action, while men more often use them to excuse their behaviour or to raise their status. Women tend to report nightmares somewhat more often than men, though older women and girls do not do so.

In relation to mystical beliefs and religion a recent survey found that women are more likely to believe in God and pray more often than men. This may be related to concern for children. Women are also more likely to have supernatural beliefs than men, to believe in astrology, haunted houses and communicating with the dead. Both sexes believe in witchcraft but men have a greater belief in UFOs. A survey in the United Kingdom in 2007 found that belief in telepathy was very strong among women and that one in four consulted their horoscope regularly and believed that horoscopes accurately predicted events. Whether this has a biological cause is not known!

Hertenstein and Keltner have researched gender and the communication of emotion via touch, and found that although it was thought that women would be able to communicate sympathy and happiness through brief touches to the arm of a stranger, whereas with men it would be to communicate anger, in an experiment testing touch in relation to different emotions, there were no gender-related differences in the communication of disgust, fear, envy, embarrassment, sadness, pride, love or gratitude. Sympathy was communicated accurately through tactile contact to the arm only in couples including at least one female; couples consisting of two men communicated sympathy at less than chance levels. Anger, in contrast, was communicated accurately only when the couple included at least one male. Women are more likely than men to perceive a touch from a strange man as unpleasant and an invasion of privacy. The more women perceive a touch from a male stranger as sexual, the less they perceive the touch as warm and friendly, whereas the reverse is true for men receiving a touch from a female stranger.

When men and women talk, in the words of Professor Robert Provine, 'Females are the leading laughers, but males are the best laugh getters.' Women laugh more than men, whereas men are more likely to make other people laugh. Gender differences in humour appreciation show that women rate neutral humour, not sexual or hostile humour, higher than do men. Men may have a better sense of humour than women – or so they claim – and there are more male comedians than female. Women are more attracted to men who make them laugh, and both sexes rate female laughter as more 'friendly' than male. In general both sexes share much of the same response to humour although they have preferences for different types of humour. Women prefer more anecdotal humour, related to their everyday lives and social situations, while men prefer more risqué subjects. You are more likely to find a man making jokes at the expense of others, and even joking about death, than a woman. Men are also more likely to use humour to cope with stressful situations. But there are many cultural differences in what is considered humorous and defining a joke will differ from country to country. But in general, it is clear that women are more attracted to men who make them laugh, and men are likely to choose women who laugh at their jokes. Personally, I find this very comforting.

9

Mathematics

Whatever women do they must do twice as well as men to be thought half as good. Luckily, this is not difficult.

Charlotte Whitton

Women are under-represented in engineering and in academic positions in research at universities, especially in maths and science. In the United Kingdom in recent years only about five per cent of working women were employed in science, engineering or technology, compared to about a third of working men. These figures have persisted despite the fact that women account for nearly half the total workforce. There is also a gender pay gap, as women in science and engineering earn less. Similarly, only ten per cent of board directorships were held by women across fifty-three companies in science and technology, with exclusively male boards still existing in about a quarter of companies. Women in science, engineering and technology are less likely to obtain permanent full-time academic positions. In the United States there are few tenured professors in science-related departments. Why are women under-represented in these subjects?

Of all differences between the sexes in cognitive abilities that might be responsible, differences in mathematical ability have received the most attention. What might be the relative roles of biological and social factors in any such differences? Both will be seen to be involved. A major factor is the choice made by women with respect to the subjects they wish to work in, and both biological and social factors are involved.

As we saw earlier, in ancient times women were the inventors of agriculture and have also been responsible for related inventions throughout history. But in mathematics and other science-related subjects women were absent from very early times, with only a few exceptions. Merit Ptah (about 2700 BC) was an early physician in ancient Egypt, and the first woman we know of by name in the history of medicine. Agnodice was the earliest midwife to be mentioned among the ancient Greeks. She was a native of Athens, where women and slaves were excluded by law from the study of medicine; she attended lectures in midwifery and gynaecology disguised in men's clothing. From its beginnings in Greece, science was male-dominated. Although women were not excluded from the study of science in ancient Greece they made few contributions. An exception was Hypatia of Alexandria (about 350–415 AD) who wrote texts on astronomy, geometry and algebra, and is remembered for several inventions, including a hydrometer for measuring the specific gravity of liquids, an instrument for predicting the position of heavenly bodies and another for distilling water.

Higher education was denied to women and when universities began in the eleventh century, all except Bologna excluded women. Women made no more progress in the Middle Ages, hardly encouraged by attitudes such as that of St Thomas Aquinas who wrote, 'A woman is mentally incapable of holding a position of authority.' But between 1650 and 1710, women made up fourteen per cent of all German astronomers, the most famous being

Maria Winkelmann, However, despite her qualifications, she worked at the Berlin Academy only in an unofficial capacity, assisting her husband who had been appointed official astronomer. When she discovered a comet in 1702, the credit went to her husband and not to her, and when he died, she was denied an opportunity to take his post, which was given to a much less qualified man. Even the Enlightenment was not very enlightened. Many believed women's work should be as wife and mother. Nevertheless Eva Ekeblad, whose discoveries included how to make flour and alcohol from potatoes, became the first female member of the Royal Swedish Academy of Sciences in 1748. Important progress in gender equality was also made by Émilie du Châtelet, who translated Newton's *Principia Mathematica*, and by Caroline Herschel the German–British astronomer who identified eight comets, and became the first woman in Britain to be paid as a scientist in 1787.

It was only in 1849, when Elizabeth Blackwell became the first certified female doctor in the United States, that there was any further significant progress. Marie Curie in 1903 became the first woman to win a Nobel Prize in physics, and added a second, in chemistry, in 1911 to become a double Nobel Prize-winner. Both prizes were for her work on radiation. In 1906 Alice Perry became the first woman in the British Isles to graduate in civil engineering. Since the beginning of the twentieth century matters have improved greatly, with major scientific contributions being made by women. The windscreen wiper, invented by Mary Anderson in 1903, allowed street cars to operate safely in the rain. Ida Henrietta Hyde was an American physiologist who in the 1930s invented the microelectrode, a small device that electrically stimulates a living cell and records its electrical activity.

Women's delayed contribution to science does not seem to reflect intellectual weakness, but rather factors such as social prejudice and women's own preferences. A recent UK government report says that there is evidence that at school, when still under the age of five, girls outperform boys on all aspects of learning and this continues up to the sixth form, though boys get more top grades. This leads to the view that social rather than biological factors must play a major role, and that the negative influence of female stereotypes is significant in excluding women from certain disciplines, for example, engineering and mathematics. In the United States women have received more college degrees than men every year since 1982, and with each year the gap widens By January 2013 it was over nine million more, according to the US Department of Education. No single factor has been found to cause sex differences in science and maths. Biological differences and the way topics are described and taught could both be involved. A very significant cause could be the influence of teaching which gives rise to a negative stereotype for female mathematicians. For example, by as early as the second grade (around the age of seven) American children expressed the cultural stereotype that maths is for boys, and elementary-school boys identified with maths more strongly than did girls. The findings suggest that the maths gender stereotype favouring boys is acquired early, and that it influences emerging self-concepts of mathematical ability before the age at which there are actual differences in mathematical achievement. Research on stereotypes in children suggests that gender identity disrupts girls' maths performance as early as five to seven years of age. Girls who are reminded of supposed sex weakness in maths abilities before being tested perform worse than those who are not told of the stereotype. Positive stereotypes, such as a message that girls are very good at maths, can improve performance.

Most elementary-school teachers in the United States and primary-school teachers in the United Kingdom are female. Beilock and her colleagues have found that when female elementary-school teachers are themselves not confident about maths, this anxiety carries negative consequences for the maths achievement of their female students. If teachers pass on the commonly held view that 'boys are good at maths, and girls are good at reading' this can have a negative effect on girls in relation to maths. An investigation by Haworth and her team, about whether the sexes differ in science performance before they make important course and career decisions at ages nine, ten and twelve, found there was no evidence for mean sex differences in science performance. At a time when adolescents are making important course choices, girls are performing just as well as boys. Other studies also found no evidence for quantitative or qualitative sex differences in science. Since girls are just as good at science as boys in early adolescence, the lack of women in scientific careers, particularly in the physical and engineering sciences, could be due to social factors.

A controversial and very different view was put by Harvard President Lawrence Summers, who in 2005 suggested that women might be under-represented in the sciences because of a lack of 'intrinsic aptitude' for science compared to men. Speaking of the finding that there are many fewer women than men at the upper end of advanced mathematical achievement, Summers said:

> It does appear that on many, many different human attributes – height, weight, propensity for criminality, overall IQ, mathematical ability, scientific ability – there is relatively clear evidence that whatever the difference in means – which can be debated – there is a difference in the standard deviation, and variability of a male and a female population. And that is true with respect to attributes that are and are not plausibly, culturally determined. If one supposes, as I think is reasonable, that if one is talking about physicists at a top 25 research university, one is not talking about people who are two standard deviations above the mean. And perhaps it's not even talking about somebody who is three standard deviations above the mean. But it's talking about people who are three and a half, four standard deviations above the mean in the one in 5,000, one in 10,000 class. Even small differences in the standard deviation will translate into very large differences in the available pool.

His statement implies that the skills required to be a good scientist are innate, and men have them to a greater degree than women. Although he did discuss the possibility that sex-related differences in socialisation or sexual discrimination during hiring or promotion could be barriers to female success in science, mathematics and engineering, but he did not rate these factors highly. The remarks caused a backlash; Summers resigned his Harvard post in 2006, partly because of this controversy.

We now know that sex influences on brain function are quite common, and the question is whether the differences in representation in science-based positions have a biological cause. The scientific literature is filled with studies of cognitive sex differences, but general intelligence is not an issue, as IQ tests have shown minimal or negligible differences between men and women. At primary school there is a small difference between boys and girls in mathematical abilities and it favours girls. There are no essential differences in primary maths abilities such as counting and doing simple arithmetic in childhood. One analysis of gender differences in over a hundred studies found that performance on a range of maths tasks was similar in girls and boys. But differences emerge later in geometry and other fields. Across diverse cultures girls are better at calculation while boys are better at mathematical problem-solving, and a male advantage appears when the maths requires more reasoning and is more spatial in nature, as in geometry and calculus. How children play may be related to sex differences in some skills related to maths and the sciences. Boys select play areas that are one and a half to three times the size of girls' play areas, and they organise the space with buildings like forts. Activities of this kind this might contribute to an ability to visualise and remember the geometric features of large-scale space.

The most obvious difference in maths ability is that men are often over-represented at extreme scores, both very high and very low, and this can make a big difference in the numbers selected for academic posts. Among those who took the SAT-M maths test at school in the United States, large sex differences were found at the highest end of the distribution. The test is designed to measure ability to solve problems, not to gauge maths knowledge, and although the questions require only basic maths and all have simple solutions, they call for considerable ingenuity. Boys are more variable in the relevant abilities and there are more boys at both the high and low extremes. But among the top one per cent of scorers there are twice as many boys as girls. There is thus a strong male preponderance for outstanding mathematicians – but there are also many very competent female mathematicians. A possible reason why males show more variability may be that the inferior parietal lobule in the brain is linked to mathematical ability and is typically larger in men, especially on the left side, than in women. When men and women with equal ability in maths are tested doing maths, there is temporal lobe activation in men but not in women. There is evidence, to be discussed, for a male advantage in some visuospatial rotation tests, and this ability may make them more skilled in engineering and science tasks that involving the manipulation of objects. This may involve the intraparietal region, the principal functions of which are related to perceptual-motor co-ordination. Einstein's brain is claimed to have been abnormally large in this region.

In the United Kingdom no significant male–female differences in ability are indicated by the numbers graduating in science subjects or by A level results in maths. In the United States girls at all grade levels now perform on a par with boys on the standardised maths tests required of all students, and the percentage of US doctorates in maths and the sciences awarded to women has climbed to thirty per cent, up from five per cent in the 1950s. In the UK girls perform as well or better in GCSE and A level science-related subjects, but form a smaller proportion of the A level entrants to most of these subjects. For example, boys made up seventy-nine per cent of all pupils taking physics in 2013and only six and a half per cent of those taking computing were girls. Even of those who had graduated in these subjects a high percentage are not using their qualifications to work in related occupations

Ability at mental rotation, where a subject is asked to compare two three-dimensional objects or shapes, and say if they are the same image or mirror images, is a prominent skill in males but it is not closely associated with performance in maths. Spatial and maths performance are largely independent of each other, but statistically significant correlations between maths and visualisation and mental rotation have been reported. Social scientists often cite gender differences in spatial abilities as a crucial factor in discouraging women from entering fields such as architecture, engineering, physics and maths. This fits well with the view that men are systemisers who are naturally at home in these subjects, whereas women are not. Another possible reason for women being less involved in maths and the physical sciences is that they have an image of these as being closely connected with 'forces'. Some images of engineering conjure up the enormous forces involved in, for example, building a skyscraper. Thus the greater physical strength of men could predispose men rather than women to enter such fields.

Probably more important is the idea that maths and physical sciences do not involve empathy. As Simon Baron-Cohen points out, this lack would alienate females.

He argues that men and women are naturally suited to different kinds of work. In *The Essential Difference* he offers the following 'scientific' careers advice:

> People with the female brain make the most wonderful counsellors, primary school teachers, nurses, carers, therapists, social workers, mediators, group facilitators or personnel staff . . . People with the male brain make the most wonderful scientists, engineers, mechanics, technicians, musicians, architects, electricians, plumbers, taxonomists, catalogists, bankers, toolmakers, programmers or even lawyers.

For Baron-Cohen the difference between the two lists reflects what he takes to be the 'essential difference' between male and female brains. The female-brain jobs make use of a capacity for empathy and communication, whereas the male-brain jobs exploit the ability to analyse complex systems. Another possibility is that women wish to avoid highly competitive fields like these. However, some feminists have argued that Baron-Cohen is trying to suggest that women are better at low-paid and menial work, to the advantage of men.

Female choice of a career can be greatly influenced by thinking about having children, taking care of elderly parents or staying geographically close to home in order to remain with their family. This isn't necessarily a female choice, as there is a clear expectation that women will do the majority of the work rearing children and taking care of the elderly. Factors such as maternity and paternity leave laws, employers' attitudes to mothers in the workplace, and society's attitudes to house-husbands will play a role, as well as basic biology. One woman scientist known to me has stated that she believes she would have been able to run her lab very successfully if she had been permitted to share the work with a close female colleague, who also had young children.

A number of factors are required to become a successful scientist and these are not well understood, but as we have seen, males show more variability which results in a greater number having the highest skills. There is also the key issue of how being a scientist is perceived by young girls and boys. They know little about what it is really like to be a scientist but undoubtedly have various images from the media. But different fields of science require different skills and attract male and female students in different ways. Biology, for example, is different from maths and physics as it relates more closely to humans. A key issue is of course how genuinely interested the individual is in science. It is important to recognise that there are many distinguished women scientists and I have had the pleasure of collaborating with several.

There is evidence for a reduction in interest by adolescents for school science, especially more for girls than for boys. Girls often have greater verbal ability than boys, and this gives them more career choices than the boys, whose strength is mainly in the sciences. Schoolchildren in the United Kingdom have to make an early choice at about age fifteen and sixteen as to whether they will specialise in the arts or science. There are many determinants in choosing a course and later a career, and both abilities and personal beliefs are thought to be important. One poll of eight- to seventeen-year-olds reported that twenty-four per cent of boys were interested in engineering versus only five per cent of girls; a survey of thirteen- to seventeen-year-olds reported that seventy-four per cent of boys were interested in computer science while only thirty-two per cent of girls were. Girls with CAH were found more likely to express interest in careers often seen as the preserve of the male, such as engineering, working in construction, or being an airline pilot.

A primary factor in women's under-representation in maths and engineering careers is thus their choice when young. They prefer, for example, to be doctors, vets, biologists, psychologists or lawyers rather than engineers or physicists. Men are very much more likely than women to apply for a job with a salary potential that is dependent on outperforming their colleagues, according to a large new study from the University of Chicago by Flory's team. 'Women shy away from competitive workplaces whereas men covet, and even thrive in competitive environments,' concluded the study, involving nearly 7,000 job-seekers in sixteen large American cities. Women tend to prefer to work with people rather than things and so prefer medicine rather than technological subjects. They now outnumber men in medical schools in the United Kingdom, and female doctors will soon outnumber male doctors. But there are concerns that they still do not occupy many senior positions. In the United States more women than men who enter graduate study in maths-related fields drop out, and fewer of those who complete doctorates apply for tenured academic positions. Women drop out of scientific careers – especially maths and physical sciences – even after becoming assistant professors at higher rates than men. This is not, apparently, to do with discrimination or ability but rather with lifestyle choices. But the fact that women are not systemisers must play an important role.

In the United States women obtain at least forty per cent of bachelor degrees in maths, biology and chemistry, but the proportion who earn bachelor degrees in physics has been unchanged at just above twenty per cent since the turn of the century. The engineering workforce remains the area of highest under-representation for women, as in 2003 only eleven per cent were women. However, in China forty per cent of engineers are now female. In computer science especially there has been a serious drop in women earning degrees. There are claims that some aspects of computing may discourage women, particularly what has been called the 'geek factor'. The image of a computer scientist sitting in front of a screen all day is not an appealing image for teenage girls who look for more sociability in the workplace. A highly relevant study by John Chen and his team of the performance of 126 female mechanical engineering students at North Carolina A&T State University in a range of classes, each emphasising different skills, found that the female students performed better on every measure than the males in all but one class. The reasons were related to the better preparation of female students for entering into engineering studies, as well as the relatively high proportion of women in mechanical engineering at the college, which provided a more supportive atmosphere for the female students. These led to greater self-confidence with respect to their skills and abilities. This is a set of objectives to be strongly encouraged.

There are some interesting figures for other occupations. The statistical breakdown of women in different occupations is quite complex, but on the science side the numbers are low. Only one in twenty of all working women is employed in any maths-, science-, engineering- or technology-based job, compared to nearly one in three of working men; and only about a third of all female graduates of working age in this area are employed in related occupations compared to half of all male graduates. There is also a 'leaky pipeline', where women in these areas leave their careers and struggle or fail to return. Of surgeons in the United States only nineteen per cent are women; the figure is just eight per cent in the United Kingdom, although forty-three per cent of NHS dentists are now female, forty per cent of UK lawyers are women and eighty per cent of UK primary-school teachers are female, while the figure for those teaching in higher education is forty-three per cent. Elsewhere women owned nearly thirty per cent of all non-farming businesses in the United States in 2002, and it is impressive that more than twenty countries currently have a woman holding office as the head of a national government.

But a large number of women in maths and science areas have reported significant sex discrimination, and overall they earn significantly less than men in physical sciences, maths and engineering, as well as in life sciences, health and teaching. Yet there is no good evidence for discrimination against women in maths or science by grant agencies, journal reviewers and search committees. Evidence shows that women in these fields do as well as men in funding and publishing when given comparable resources to men. The fact that they are often not given the same resources may be due to family commitments and lifestyle choices they have made rather than sexual discrimination.

Women's limited numbers in the physical sciences thus reflect choice influenced by empathy. They still have primary responsibility for child care, and may need to work fewer or more flexible hours than men. There is also a view that being successful in careers such as maths or engineering may make women less attractive as marriage partners. Girls' lack of maths and science motivation is positively associated with their mothers and peer-group support. Women appear to favour a more communal, holistic perspective of life relating to family and friendships, with less time devoted to a career. Although they are at least as able as other workers, mothers may need to prioritise short, flexible working weeks and part-time work. Men, by contrast, can focus much more time on their careers.

There is evidence that single-mindedness, an attribute important for maths and science, is particularly common in males. The web pages of the Smithsonian Institute in Washington suggest that those working on review problems number about thirty women and 125 men. Again, there is evidence that men do better at producing significant research. Recently among student members of the British Psychological Society, there were 5,806 women to 945 men, and among graduate psychologists, 23,324 women to 8,592 men. Of those who practise as chartered psychologists, there were 7,369 women to 4,402 men. Yet among Fellows of the Society, honoured largely for their research, there are 428 men to only 106 women. This may reflect commitment.

A provocative suggestion is that men may have realised that a reputation for being no good at housework can some-times work to their advantage, as women will then do it all, leaving them more time to concentrate on their careers.

10
Skills

I hate women because they always know where things are.

Voltaire

There are no significant variations in general intelligence between the sexes, but the areas in the brain where they generate their intelligence are significantly different. Melissa Hines has noted that there are many differences in brain function during equivalent performances by the two sexes. For instance, men and women show different patterns of asymmetry of brain function when performing certain tasks involving listening, despite showing no sex difference in task performance. But there are differences in certain skills and it has long been held that as mentioned earlier, men have more grey matter and women more white matter, possibly related to some skills. Human evolution has resulted, it seems, in two similar but different types of brains with equally intelligent behaviour, but their characteristics can be mixed. There can be women with features of a 'male brain' and vice versa.

As Uta Frith, the distinguished neuroscientist, has pointed out to me, one can achieve the same performance on a test by entirely different means. Equal performance on, for example, an IQ test does not allow one to say each subject has achieved it in the same way. It seems to her quite likely that in the future many differences in male and female brains will be found, and these may not relate to performance levels at all. One needs to realise that brains combine 'male' and 'female' properties.

Information-processing centres in the brain are based on grey matter while the connections between these centres are made by white matter. These differences could explain why some men are very good at some tasks, such as mathematics, which require more local information processing, while some women tend to excel at language which it is claimed uses white matter to integrate and assimilate information from the various grey-matter regions in the brain that are required for language. Most grey-matter regions and white-matter regions involved with intellectual activity in women are in the frontal lobes, compared to very much less in males, where the relevant grey matter is distributed throughout more of the brain. The fact that if a woman receives an injury to the front of her brain, she may suffer more cognitive damage than a man injured in the same region fits nicely with these differences. Differences between the left and right hemispheres of the brain could explain why men tend to deal with problem solving by using the left half of the brain and taking a problem-solving approach akin to mathematics, while women typically solve problems using various grey-matter areas that are linked to language and are more aware of their feelings. There are also structural sex differences in areas of the human visual cortex which are related to the detection of motion. These might give males the better skills required for spatial targeting, and have possibly evolved from ancient hunting skills. Women, on the other hand, are better at detecting colour changes. It seems, as we have just seen, that personal choice rather than intrinsic ability underlies apparent sex differences in science-based careers. What about other skills, such as the performance of skilled motor tasks? One too often comes across beliefs in stereotypes, such as women are better at multitasking but are bad drivers and poor at parking. We will examine the evidence for these assertions.

In the United States a variety of skills are claimed by Diprete and Jennings to be important for success in primary school and girls have been found to have a considerable early lead over boys in these social and behavioural skills, which continues during their schooling. Moreover this female advantage persists when they go to college, and is considered to be the single most important factor underlying the significant lead that women have over men in rates of college completion.

There are a number of quite well established sex differences in skills related to cognitive functioning. Women are best at tasks involving verbal skills and episodic memory – recalling specific events – face recognition, perceptual speed and accuracy, and fine motor skills, while men typically have stronger spatial abilities, and are better able to process shapes and how the dynamics of these shapes change . . . We will return to verbal skills in the next chapter.

We have already mentioned mental rotation tests, where subjects have to compare three-dimensional objects or shapes in their heads. Men do very well with this test but most women struggle. Women have a thicker parietal region of the brain, which slows down their ability to mentally rotate objects. Research has shown that the sex difference in this ability is present in babies as young as five months, thus ruling out any idea that these abilities are determined by environment. Male hormones play a key role, and Berenbaum and his fellow researchers have demonstrated that women with CAH have greater spatial ability than their sisters throughout life. In addition women with a male twin usually have greater spatial ability than those with a female twin, as the male twin provides more testosterone in the womb. A well-established difference is that although men score higher than women on tests of spatial perception and orientation, women score higher than men on tests of memory for object locations, such as where a tin of soup is in a large supermarket. Explanations for such differences have focused on various causal factors, like biological evolution and the effects of prenatal and post-pubertal sex steroid hormones. Evolution has very likely played a key role. The paper by Lippa, Collaer and Peters suggests that some spatial abilities could have evolved from ancestral men's devotion to tracking, hunting, targeting and projectile throwing which no doubt favoured the development of three-dimension visualisation skills and required the ability to visually track and target moving objects. By contrast, ancestral women's focusing on foraging may have favoured the development of accurate memory for object locations and skill in locating foraging sites in relation to geographic landmarks.

Lippa's paper quotes data from a BBC Internet study showed that, across fifty-three nations, men outperformed women on two visuospatial tasks. The extent of the male advantage tended to be positively associated with nations' gender equality and economic development. As mentioned earlier, sex differences are usually larger in gender-egalitarian, economically advanced nations. Men are better not only at some visuospatial tasks, but also in auditory-spatial tasks. There are sex differences in extracting specific information of interest from a situation composed of multiple sound sources – the cocktail party as studied by Zündorf and his team at Tübingen University. Participants were asked to listen to sounds and determine the location of the sound source by pointing towards it or by naming the exact position. When sounds were presented one at a time, both men and women accomplished the task with great accuracy, but when several sounds were presented simultaneously and participants had to focus on and localise only one sound, women found the second task much more difficult than men.

Women do better on precision manual tasks involving fine motor co-ordination, such the assembly of circuit boards in a factory, and this may be a result of foraging skills that evolved a long time ago. This ability of women to perform better at fine motor activities may be due in part to their being more patient. Women also may do better on tactile tests, partly because they tend to have smaller hands.

While some studies have found that women with CAH resemble men by being better at mental rotation, other studies have found that male subjects with CAH show a paradoxically reduced skill for mental rotations. This is not understood. Most studies have found no differences in individuals with CAH for tasks such as verbal fluency or perceptual speed, at which women usually do better, though one study suggests they are less adept than other women at fine motor performance tasks. It is possible that exposure to testosterone in the womb has a greater impact on sexually linked motor abilities such as targeting in men and fine motor control in women. Girls and women with CAH are better at targeting – more like men. In women the menstrual cycle alters the performance of tasks related to motor and spatial ability, with high scores during the menstrual phase and low scores during the mid-luteal phase when higher levels of progesterone are found.

Experience can alter mental rotation skills. Children were given mental rotation tests before and after some training in how to do them and the initial gender differences disappeared following this training. More importantly, changes in the ability to do the rotation test can be brought about by information relating to which sex is best at it. Once again, stereotyping can affect performance. If women are told before a rotation test that men typically perform better, or that the task is linked with jobs such as aviation and engineering rather than jobs typically associated with women, they will perform worse. If they are told the opposite, women will do better. In another study those in the negative stereotype group showed greater activation of the amygdala, which is associated with fear and other negative emotions.

That differences in spatial reasoning may not be innate comes from a study in primitive societies. In one patrilineal primitive society, men outperformed women on a spatial reasoning task but that difference vanished when the test was given to matrilineal tribe members where women hold more power – then there was simply no significant difference between the sexes.

Another important spatial ability is differentiation of left and right. Confusion is common even in healthy adults. Women are more prone to such errors than men, as confirmed by tests asking subjects to identify the left or right hand in a drawing. Men are less likely to have to think about which way to turn a screw. They outperform women in overall spatial ability, but the sex differences in spatial memory are not the same across all forms of spatial representation. Women have been found to excel at landmark memory images of differing types of structures. They also have an advantage in tasks related to declarative memory, the retrieval of long-term memories of specific events and facts, so may have an advantage in general knowledge. Hormones are once again involved and oestrogens may affect declarative memory, which is linked to the white-matter microstructure underlying the parahippocampus. Mendrek and her colleagues found an overall positive correlation between a normal range of circulating testosterone and successful visuospatial abilities in men, but men with both extremely low and extremely high levels of testosterone are associated with poor performance in this area – very puzzling.

Women have a very good episodic memory, a memory of autobiographical events in their lives including words or things that happened on a daily basis, while men are better than women at recalling non-verbal visual information such as remembering the way out of a wood. Men have been found to be better than women at finding their way when far from a goal. Women, on the other hand have been shown by Hassan and Rahman to perform better than men, on average, on tests of object-location memory. Specifically, women outperform men in location recall of objects that have exchanged places or shifted position, or when new objects are added to a previously studied array. They are thus better at remembering where they left the car keys.

In addition, as we have seen, women are better than men at remembering faces, especially of other women –

perhaps because they give more attention to female than to male faces whereas men show no such own-gender bias. Women do better than men on all measures of facial recognition and these female advantages have been reported cross-culturally. Tests in which subjects must rapidly identify matching items are done better by women. Homosexual men are better at object-location memory than heterosexual men, and thus more like women, but on tests of spatial cognition they are less skilled than heterosexual men.

Tests on the ability to work out new spatial information from memory were based on experiments in which the subjects had to study line drawings of shapes linked either by straight or meandering lines, and then remember the straight-line distances. Men performed better. While women tend to rely on landmarks when they learn a route, either from a map or from direct experience, men are more likely to be aware of the compass direction in which they are travelling, and when using a map, the men made fewer mistakes in getting to a destination. So maybe the old argument that men are better at map-reading than women does have some validity.

There may be an evolution-based origin for the sex differences in spatial navigation as sex-related differences in this are seen also in other species. Hormones may again be playing a role, since if a female rat is injected at birth with testosterone, she learns how to get out of a maze faster. The right visual-association areas of the brain have been found to be more activated in men when doing spatial tasks in a three-dimensional virtual maze. Women, however, used additional regions on both sides of the brain for the more difficult spatial tasks. This difference was also found when doing mental rotation and numeric calculation (see Ira Epstein *What Every Man Really Needs*). Groups of young teenage boys and teenage girls attempting to find their way out of a maze again illustrate some of these sex differences. Whereas boys generally appoint a leader who has demonstrated his skills, and tend to search and explore the maze using scouts while remaining far apart from each other, girls tend to work in groups to explore the maze together without establishing a clear leader. So boys tend to use a command structure where girls use equal relationships and group co-operation in trying to discover how to find their way out.

Male superiority emerges on spatial tasks by the age of four, and this includes spatial tasks that measure accuracy. For example, large differences have been found with judgements about moving objects, such as studies that used 'time to collision' estimates. The tests are about the ability to judge when a moving object will arrive at a target, and the evidence is that males do much better at doing this. This may be related to sport, as testosterone levels from puberty are decisive with respect to physical performance by both male and female athletes in high-ranked sports. There is also a significant sex difference in throwing accuracy, which reflects the advantage of men in targeting abilities. There are also differences in spatial perception which requires targeting, such as hitting a target with a ball. Here again, men do better. Women are also worse than men at predicting the level of water when a half-filled bottle of water is tilted, but this may be due to a poorer understanding of basic physics.

Women have less upper-body strength than men, averaging only eighty per cent of that of men of identical weight. Sex differences in the ability to throw a ball far and accurately develop by the age of three. Men are also better at darts, but their steadiness makes women good at archery. There are currently no sports in which women are better than men, probably since men are bigger, stronger and faster and have better spatial and motor skills. But women are improving so rapidly that they may catch up even in competitions like the marathon, where the fastest woman is now just fifteen minutes behind the male champion. In the Olympics of 2156, women's times may beat men's in the hundred-metre sprint as the time difference between men's and women's speed has been getting less. There is also a possibility that stronger competition between women may lead to more practice and more intense play, which may in turn translate into superior skills and scores. Sex differences in motor abilities also have been related to testosterone exposure in the embryo and women with CAH showed more accuracy in throwing balls and darts at targets, and this finding was not related to an increase in the strength of their muscles.

The ability to conceptualise and manipulate tools in a complex manner is a distinguishing characteristic of humans, and was a major milestone in human evolution. Boys are more socialised with respect to technology as shown by three-year-old children during play sessions when they had to retrieve a toy which was out of reach using just one of six tools. The boys were more interested in objects and tried using the tools to get the toy. Using tools is a motor act and may involve mental rotation skills and boys and girls recruit different brain regions for tool use. Imitation can be important when learning new motor skills. Nichola Rice Cohen and her co-authors have shown that imitation is of critical importance in learning new skills. The imitation of hand movements involves a region of the brain in the left-hemisphere system known as the praxis system. In an experiment subjects viewed a sequence of hand gestures and then had to reproduce them from memory. The movement of the subjects' hands were recorded by a special glove and analysed with a technique that identifies spatial and time differences between the original hand movements and the subsequent imitation. Results confirm a female advantage, as men left out more gestures from their remembered performance than women do. This difference reflects a female advantage when it comes to motor planning at the level of individual items. This female advantage is consistent with experiments with preschool children in which girls imitated gestures more correctly.

A recent official report by the United Kingdom Driving Standards Agency provides some evidence for the idea that women are not very good at parking a car, and are twice as likely as men to fail their driving test when trying to parallel park. Women needed an average of fifty-two hours of tuition to pass the test whereas men required, on average, just thirty-six. However, surveillance of British car parks has shown that while women may take longer to park, they are more likely to leave their vehicles correctly in the middle of a bay.

In relation to music and related motor skills, tests have revealed no significant differences between the sexes, but in a rhythmic ability test girls outperformed boys in four of the six movements. There may be some evidence that testosterone levels are related to talents in musical composition, as there are few famous women composers.

In chess there has never been a female world champion, and only one per cent of Grand Masters are women. But the under-representation of women at the top end in chess is what would be expected from the much greater number of men who participate in the game.

Men and women are thought to approach problems with different methods even if they have similar goals. They can solve many problems equally well, but their approach and the processes they use are often reported to be quite different. For most women, sharing and discussing a problem, as we have already seen, presents an opportunity for women to get to know better the person with whom they are collaborating. Women are also usually reported to be more concerned about how problems are solved than with merely solving the problem itself. They are said to think intuitively, and to be able to handle many sources of information at a time, but this can have the drawback of causing them to be overwhelmed by the complexities of a problem. They may also have difficulty in separating their personal experience from the problem in hand. Men, on the other hand, tend to focus on one problem at a time or on a limited number of problems, and apparently are more able to distance themselves from problems and minimise the complexity that may exist. But it is also claimed that in minimising difficulties they sometimes fail to appreciate subtleties that can be crucial for a successful solution.

When men, it is claimed, think about a problem or something that interests them, they keep their mind on it all the time and nothing else then exists for them except for what is related to that issue. On the other hand, women tend to do the opposite, and think more generally in trying to connect things up with other things. Do not try to talk to a man about something you feel is important while he is concentrating his attention on a specific problem or even on a football match, as it is unlikely you will get a useful response. As we have seen, men tend to process thinking better in the left hemisphere of the brain, while women tend to process equally well between the two hemispheres, and this may be why they approach problem-solving in different ways. Men rely more on the left brain to solve a problem one step at a time. Women make more use of the right hemisphere because they have more efficient access to both sides of the brain. This may help them to focus on more than one problem at one time, and they often prefer to solve problems using many simultaneous mental activities, which could help with multitasking. A lot of parents have observed how young girls can take part in a conversation among several other girls who may be discussing as many as three subjects at once. This leads me naturally to questions of language ability.

Sweden has made serious efforts to eliminate gender discrimination, and in some schools the most popular toys for both boys and girls is a set of naked dolls with no signs of gender. There is also use of a gender neutral pronoun, *des*, so no 'him' or 'her'. What effect will this have on their current and later behaviour?

11

Language

Men and women belong to different species and
communication between them is still in its infancy.
Bill Cosby

A widely held belief is that language and communication
are more important to women than to men, and that
women also talk more than men and are better with words.
It is also claimed that whereas women talk more about
people, relationships and feelings, men talk about things
and facts. Such ideas have repercussions on our beliefs and
our actions, but they are essentially myths for which there
is relatively little evidence. Research has shown that the
overall difference between men and women with relation
to language is usually small or close to zero. The myths
of 'Mars and Venus' provide no exception to that rule.
The idea that men are no good at talking seems somewhat
amusing when viewed as a useful myth that might exempt
men from something they would rather leave to women.

The two chief areas of the human brain responsible
for language are Broca's area, which is partly responsible
for the production of language, for example when putting
together sentences, and Wernicke's area, which is partly
responsible for processing other people's sentences.
Both regions are claimed to be proportionately larger in
women. Similarly the hippocampus, a structure involved
in learning, memory and emotion and in translating
memories into words, is said to be larger and more active

in women. It is also oestrogen-sensitive. There is, however, one report of a test on verbal fluency from the Université de Montpellier in which men showed greater activation than women in these classical language regions of the brain, which shows that the significance of these claims is far from clear.

Women's brains have been reported to be more bilaterally organised for language, whereas men show greater left-hemisphere dominance, but this is no longer generally accepted. A recent large, multi-task investigation of sex differences both in structural asymmetries and in the lateralisation of function in reading found little evidence for sex differences on any behavioural measure. Sex differences in the brain in relation to language accounted for two per cent or less of the individual variation in asymmetry. But it has been widely held that language is more left-lateralised in males.

In contrast to maths and mental rotation, where men have the advantage, women are widely claimed to have an advantage over them in verbal skills, such as vocabulary use, reading, comprehension, essay-writing, verbal fluency, verbal learning and memory – but the evidence is mixed. Some research has shown a greater prevalence of pre-verbal skills among small girls, with female dominance in adults in language ability including grammar, spelling, and remembering lists of unrelated words. But the literature on sex differences in language can be no less confusing than that related to maths and science. In 1988 a review of numerous studies showed that female superiority was so slight as to be meaningless. There was no evidence of significant differences between the sexes in any component of verbal processing. However, a study of 165 papers showed that the gender of the author did have significance on the findings. If the first author was female, the article was significantly more likely to report that females had a language advantage than if the first author was male.

The background for much research on this topic is an evolutionary explanation based again on the hunter-gatherer hypothesis. This proposes, as we have already seen, that sex differences in cognitive abilities derive from the way that obtaining food was divided between the sexes in prehistoric humans. Men did most of the hunting, while women did most of the gathering of food such as wild berries. Since men explored larger territories during hunting than women did when gathering, this led to the evolution of better navigational skills in men. On the other hand, women staying at home in social groups would have interacted socially to a greater degree and used language more. Thus evolutionary theory predicts that women would have become better than men in the use of language. The hunter-gatherer hypothesis also suggests that their different roles forced men to be more aggressive and self-assertive in speech, and women to be more nurturing and empathetic in their use of language.

The auditory system, closely linked to language, shows sex differences which are probably dependent on exposure to male hormones during prenatal development. A specific sound, considered to be related to the amplification function of the cochlea, is generated from within the inner ear of newborn male infants. Their cochleas are weaker than those of newborn females, and these differences persist throughout life.

Significant language differences have been found in children. Female infants from two to twelve months old have been found to cry more in response to pain, and at a higher pitch, despite there being no evident sex differences in pain thresholds. Girls from a very early age tend to develop language more quickly than boys and acquire a larger vocabulary from as early as sixteen months. Significant differences in vocabulary have been found in both one- and two-year-olds, both in understanding language and speaking it. But although girls scored higher than boys, the differences were quite small, only about two per cent. A study of more than 3,000 two-year-old twins found that girls scored more highly than boys on both verbal and non-verbal cognitive ability, but once again the gap was tiny, a mere three per cent of the overall range. Nevertheless language performance is generally better among girls than among boys as young as two or three, though the difference largely disappears by the time they are six. There is evidence that girls have consistently outperformed boys in written work at school, though these differences become less apparent in adults. Acquiring language early seems to be part of a general, early developmental gap between the sexes, shown also in non-verbal performance.

Ozçalişkan and Goldin-Meadow made an investigation into possible sex differences in children's gestures as their language developed, and observed girls and boys every four months as they progressed from one-word to multiword speech. Boys not only produced speech combinations like 'drink juice' three months later than girls, they were also three months later in producing gestures together with speech, such as saying 'eat' while pointing at a biscuit. This was evidence that boys are likely also to lag behind girls in sentence construction; and boys were typically found to produce their first multiword sentences later than girls.

Certain language-related problems show clearer sex differences, including disorders such as stuttering, dyslexia and autism. The incidence of dyslexia and reading difficulties is higher in boys, which may be related to the greater variance in their reading performance. Men stutter more than women and there are differences in their brains related to stuttering.

Burman, Bitan and Booth found gender differences in the brain in relation to language. Girls aged nine to fifteen showed stronger responses in the language areas of the brain than did boys of the same age. The information content of some tasks was found to activate girls' language areas associated with abstract thinking, and for some of these areas the degree of activation correlated positively with performance accuracy. But reliable performance by boys when reading depended on how hard their visual areas worked. Only when hearing words, rather than reading them, did boys' performance improve. So boys have a more sensory approach than girls, but this difference may vanish by adulthood.

There are very few reports of sex differences in language-related activity in the brain based on adults matched for age and performance. Allendorfer and her co-authors showed that men and women exhibit similar brain activation during verb-generation testing. In this task the subject hears a noun and is visually instructed to think of verbs associated with the noun and then to say them. This observation is consistent with previous studies showing similar language-related activation in both sexes. Taken together these results suggested that men and women activate a similar network of brain regions during language processing, but may differ somewhat in the emotional and cognitive control strategies they employ.

In general, then, language differences due to sex are very small in almost every study. In spelling, however, there is a greater difference although it is still only moderate. Spelling scores have been correlated with grey-matter volume in the right superior frontal gyrus in women to a larger degree than in men, and they are generally better at it. Girls consistently outperform boys on tests of reading comprehension. They like to read more than boys, and more girls than boys rate themselves as confident readers. When learning similar-sounding new words, women consistently do better than men. We have already seen that women have a more efficient declarative memory system – that is, they treat memories like facts which can be consciously recalled. When subjects were asked to recall word lists in one study, girls consistently outperformed boys and the way boys and girls approached the list-learning task was found to be significantly different. Girls were more likely than boys to organise the list actively on the basis of the meanings of the words, while boys tended to recall the words in the same order in which they were presented. Accuracy of performance for boys was affected by how the words were presented, as visual presentation allowed greater recall than just hearing the word spoken. Sex differences in brain-behaviour correlations must reflect differences in the nature of the processing required.

Women have also been shown in some studies to out-perform men in semantic tasks such as tests of verbal fluency or finding synonyms. But it has also been claimed that sex is not a significant predictor in larger samples, appropriately controlled for age and education. Girls are claimed to show greater fluency as measured by the number of words spoken in a fixed time. Tests of verbal fluency involve asking subjects to come up with as many words as they can in a minute. Sometimes these words all have to begin with the same letter or sometimes they are words from within a specific category, for instance 'animals'. In these verbal tests women tend to perform at a very slightly higher level than men on most of them, but the sex difference is so small that it is virtually zero. A number of studies have looked at verbal skills in subjects with CAH, and included verbal fluency tests, but found no differences between them and non-CAH controls. However, there is evidence that at eighteen and twenty-four months, testosterone in the embryo does affect the size of vocabulary a child has; the less testosterone, the better the language skills. This may help to explain the early advantage that girls possess in acquiring speech.

Differences in language use are a complex subject. One early theory was that different ways of using or understanding language are actually displays of different power in society and that women tend to use language that echoes and helps to reinforce their subordinate role. But another more recent study found only small gender differences when it compared the use of language by men and women in both spoken and written texts. Thoughts and emotions figured higher in the list of words women used, and they were more likely to discuss people and what they were doing. For men language was more likely to be used as something to describe more concrete external events and subjects, such as jobs, money and sports, and to include numbers and swear words. Women are more expressive than men in both positive and negative emotions. It has been found that they do tend to talk about themselves and are more likely than men to include relationships in their descriptions of their ideal self and what they don't like about themselves. Men are more likely to describe themselves in terms of differences from others, rather than their connections with them.

There are no real differences between men and women when they speak about sexuality, anger or time, and in their use of 'we'. Deborah Tannen, a linguistics professor, observed that females use conversation to negotiate closeness and intimacy, so being 'best friends' means sitting and talking. For males conversation is the way you negotiate your status in the group and talk is used to preserve your independence. Whereas women may get together to chat and pursue friendships, for men undertaking activities together is the most important thing.

A surprising finding is that distinctive words, syntax, being colloquial, repetition, subordination and other features of written text can expose the gender of an author. Confirmation of the gender of a writer emerges readily from definite clues. An analysis of 14,000 text files from seventy separate studies found that while men referred more to the properties of objects and impersonal topics, women used more words related to psychological and social topics. A study revealed numerous 'psycholinguistic' factors in written material are gender-specific. A computer algorithm has been developed which can detect gender quite reliably from written texts. It makes some basic assumptions as outlined above, such as that men talk more about objects and women more about relationships. It found that women tend to use more pronouns, words like 'I' 'you' and 'we,' while men use articles that qualify nouns such as 'a' 'the' or 'that', or quantify them by using numbers.

Women currently enjoy less success than men in creative writing. In the United States eighty-three per cent of the authors reviewed in recent issues of the *New York Review of Books* were male, and the same statistic was true of reviewers. The great majority of theatre productions are written by men. In the 2008–9 Broadway season only about thirteen per cent of the plays were written by women although there are many distinguished female novelists. It is possible that these differences have a social basis, but the lack of empathy involved in writing plays may make it less attractive to women.

Over the years studies have been carried out in the United States and the United Kingdom as to how boys and girls differ in the way they use language. Perhaps unsurprisingly, they have found that the language that boys use in conversation is often more aggressive and that they interrupt more often than girls. Male contradiction starts early! On the other hand, girls are more co-operative in speech, more friendly in approach, and tend to interrupt more positively. These finding are in line with the genetic finding of greater empathy in women, and aggression in men. Tenenbaum, Ford and Alkhedairy studied girls and boys describing a wordless picture book in mixed or same-gender pairs. In mixed as well as same-gender pairs, girls used more emotional explanations than did boys, and they also used more emotional labels. In same-gender pairs girls used a higher proportion of speech that promoted collaboration than did boys, but they performed similarly when in mixed-gender pairs. The findings support contextual models of gender and suggest that speaker as well as partner gender can influence emotion expression and conversational style. Findings of Barnes,

Zimmerman, West and others showed that women often use minimal responses and encouraging noises such as 'mm-hm' and 'yeah' showing that they are collaborating in a conversation. Men use these responses less often, and when they do it is usually to show that they agree. But women will often ask a question rhetorically or as a means of drawing the other person into conversation or of getting attention.

Women use questions more frequently, while men tend to change the subject more often. It appears that women may be better listeners than men because they find listening more important than men do and because it gives them added status as a confidante. The importance women give to listening is reflected by their making fewer interruptions that disrupt conversation with unrelated topics, and by their more frequent use of minimal responses. Men interrupt far more frequently by introducing non-related topics. The interpretation of what someone is saying is, of course, influenced by its social context, such as the speaker's identity, according to van den Brink's team. If their ideas about the character of a speaker conflict with what the speaker is saying, women are much more likely than men to show specific brain reactivity. This sex difference in social information processing can be explained by a specific cognitive trait, women's ability to empathise. Individuals, mainly women, who empathise a lot show larger effects in the brain's normal response to words and other meaningful socially relevant information.

There are some linguistic skills in which girls and women have been found to be superior, such as how they pronounce words, grammar and the use of longer and more complex sentences. Havy, Bertoncini and Nazzi have reported differences in word learning, with girls being better learners than boys, as well as being less sensitive to the difficulties involved in learning. Girls seem to be capable of maintaining attention throughout a word-learning task, whereas boys display evidence of tiredness, which suggests that it is a more demanding process for them.

There is also a quite common view that women like to talk but men prefer action to words. One wisecrack is, 'I haven't spoken to my wife for three months. I don't like to interrupt her.' It is widely believed that women talk far more than men and that men tend to be generally less communicative. There is also the belief that men and women communicate differently; these stereotypes have become quite generally accepted in clichés such as 'Men never listen' and 'Women find it easier to talk about their feelings'. The idea that men and women 'speak different languages' has become deeply ingrained in our culture and is exemplified by books such as the one by Deborah Cameron about 'Mars and Venus' which accept it as an article of faith. On the contrary, that men and women differ fundamentally in the way they use language to communicate is a myth, albeit a myth from which we seem to derive some amusement. In 2006, Louann Brizendine claimed in her book *The Female Brain* that women on average utter 20,000 words a day, while men on average utter only 7,000. This confirmed the popular belief that women are not only more talkative, but almost three times more so. One person who found this impossible to believe was a professor of phonetics, Mark Liberman. He concluded that no one had ever done a study counting the words produced by a sample of men and women in the course of a single day. The figures were part of the myth. Fortunately a scientific study by Mehl and his group then found that women and men both spoke about 16,000 words per day. How many words are used does depend on the subject being talked about. The number of words used by male and female subjects during the free recall of emotional and neutral stimuli showed that women used more words compared to men in the recollection of emotional stimuli, and in describing both emotional and neutral stories.

Deborah Cameron has written that some studies have found that women talk more in interactions with partners and family members in the home, as this is a woman's domain. This may be related to the claim that in formal and public contexts, high-status speakers talk more than low-status ones. In informal contexts where status is not an issue, the commonest finding is that the two sexes contribute about equally. So why does this myth that women talk more than men persist, she asks. It possibly does so because it is based on long-established social prejudice. There are claims that women tend to communicate more effectively than men, using their abilities in deciphering non-verbal cues such as tone, emotion and facial expression as well as speech itself. Men, who have less empathy, find it more difficult to interpret these non-verbal clues and tend to rely on the spoken word alone. Such differences could help explain why men and women sometimes have difficulty communicating, and why male friendships are often based on common outside interests such as football or cars, and are different from more emotional and intuitive friendships between women. Whether such differences are cultural or biological is not known.

Men and women differ somewhat in understanding

the meaning of someone else's remarks, as the normal female brain responds earlier and more widely to words and other signals. Women identify the emotional nature of a sentence, such as sad or happy, earlier than men. Individuals with higher empathising abilities, mainly women, pick up on details about the speaker, such as their demeanour or status, while taking on board what is actually being said, thus making use of the social aspects of language. So if they are told 'I'd rather you left' they are probably able to assess if their partner really means the words he is saying. However, in circumstances where language is totally explicit such as 'Get out of my house!', sex-based differences unsurprisingly disappear.

12

Health

There is one thing women can never take away
from men. We die sooner.

P. J. O'Rourke

Among the most dreaded fatal diseases are several that
involve anatomically sex-specific structures as sites for
cancer: the male prostate and testis and the female breast,
uterine cervix and ovaries. But there are many other
diseases that for various reasons are strongly associated
with one or the other sex. Diseases that are more common
in women include certain mental illnesses like depression,
eating disorders such as anorexia nervosa and bulimia,
panic attacks and phobias. Also more common in
women are autoimmune diseases like multiple sclerosis
and rheumatoid arthritis, chronic fatigue syndrome,
and osteoporosis. There are higher rates of acute stress
disorder and post-traumatic stress disorder in women than
in men following exposure to trauma among civilians. In
general females suffer more from internalising disorders
such as depression, which result in them having difficulty
functioning and more disabling long-term conditions
such as arthritis, whereas males have more externalising
illnesses like heart attacks and stroke, which can lead to
physical disability and even death. Other diseases more
common in men include neurological disorders such as
Parkinson's disease, autism, attention deficit syndrome,
dyslexia, and addictions such as alcohol and drug abuse.

Biological differences play a key role but lifestyle contributes as well to the differing incidence and manifestations of illness in men and women. There are explanations that highlight the influence of biological sex hormones and it has been proposed that oestrogen helps to protect pre-menopausal women from cardiovascular disease. But oestrogen has also been identified as a protective factor against diseases more prevalent in men, so its rapid decline after menopause may eliminate a woman's advantage. Differences in the biochemistry of man and woman must also play a role, significant biochemical differences having been reported for 102 out of 131 key compounds. Since men have a higher incidence of cardiovascular events than women of similar ages, this led to the belief that testosterone is a risk factor for this illness in men. But this hypothesis is no longer tenable according to Guarner-Lans and his associates, as low testosterone levels are associated with multiple sclerosis, diabetes mellitus, heart disease and erectile dysfunction.

Several genetic illnesses are more common in males because of their having only one X chromosome. If a gene does not work properly on one of a female's X chromosomes, then the spare gene on her paired X chromosome can compensate. The absence of a spare X chromosome explains why certain disorders, known as X-linked disorders, are more common in men. For example, more boys and men suffer from autism than do girls and women – thus the condition probably involves genes on the X chromosome, of which there are over a thousand. Another example is haemophilia A, a bleeding disorder which affects only males and which is especially famous because of its prevalence in intermarrying European royalty. Duchenne muscular dystrophy, a form of muscle degeneration, is another X-linked male disease.

Abnormalities may be caused by variation in the number of sex chromosomes. Klinefelter syndrome, which gives males an extra X chromosome, is a cause of male infertility. It also results in small testes, a body with abnormally long limbs, wide hips which are equal to the size of the shoulders and sometimes abnormal breast development. Turner syndrome in females results in a failure to develop normal female sexual characteristics and is the result of having just one X chromosome instead of the normal two, with physical abnormalities such as short stature, swollen hands and feet and a webbed neck. Women with Turner syndrome are typically sterile because their ovaries do not function and they have no menstrual cycle. Other health problems are also frequently involved, including congenital heart disease, and cognitive deficiencies.

Although there is no doubt that biological factors contribute to gender differences in health, there is also general agreement that these do not explain all the differences between men and women. Social factors are involved as well. Women are more likely to use preventive health services, whereas a typical man will go for years without ever seeing a doctor because, unlike women, he is not trained to think about the early detection of disease. Unhealthy behaviour can also play a role. Needham and Hill have observed that men have generally been more likely to smoke and to be heavy drinkers, which may explain the male excess of kidney and liver disease. Then there are some environmental hazards associated with male-dominated occupations, such as mining and construction, which may contribute to cancer and respiratory disease, while jobs that involve repetitive tasks such as data entry or garment stitching are more commonly held by women and may give them a higher likelihood of arthritis. Women may contract infectious diseases more frequently than men because they have more contact with children. In a study in the United States it was found that each person carried an average of about 150 species of bacteria on their hands, but that women carried twice the number of species as men did, even though women wash their hands more often and are more sanitary. The reason for the difference is not known.

We now look at mental illnesses. Clinical depression is twice as common in women as in men, and is predicted to be the second leading cause of global disability by 2020. The lifetime risk of major depression in women is about twenty to twenty-six per cent, compared to about eight to twelve per cent for men. This risk is unrelated to race or ethnicity. A symptom of severe depression is total negativity about one's life. But the symptoms in men and women can be radically different; for example, men tend to blame others while women have a tendency to self-blame. Given that depressive disorders in women tend to be more prevalent after puberty and are more frequent in their reproductive years, between the ages of twenty-five and forty-four, hormonal factors probably play a role. Bereaved women are also more likely to suffer from depression. The underlying molecular mechanisms responsible for heightened female vulnerability are not known. Postnatal depression tends to develop within four to six weeks after childbirth, although in some cases it may not develop for several month and hormonal changes are not the only explanation. Menopause may lead to depression. Depressed women performed significantly worse on cognitive tests than depressed men. The female predominance in psychological disorders diminishes with age and with regard to treatment, and as we have seen, women seek help more often than men.

A study by Maguen and her colleagues of the mental health of 330,000 United States veterans returning from conflicts in Iraq and Afghanistan found that female veterans were more likely to be diagnosed with depression, and at an earlier age, than their male colleagues. Male veterans were found to be more prone to suffer from post-traumatic stress disorder and alcohol abuse.

Bipolar disorder, also known as manic depression, involves depression and mania alternating, sometimes rapidly, and develops in men and women in about equal numbers, but there are gender differences in how the illness affects them. Men typically develop bipolar disorder earlier than women and suffer more severe symptoms, particularly in the manic phases. They are more likely to display manic symptoms such as, for example, going on spending sprees. Women with bipolar disorder tend to have fewer manic episodes and more depressive ones than men do.

Suicidal behaviour linked to depression shows a clear sex difference for both fatal and non-fatal suicidal attempts. Females contemplate suicide more frequently than males, but deaths from suicide are typically higher for males. Women who attempt suicide tend to choose drug overdose or self-wounding as the suicide method, while men more frequently use hanging or asphyxia. Women are more frequently found to attempt suicide rather than to actually commit it, whereas men are more likely to complete their attempts. Older men have a higher risk of suicide than older women.

Women have a higher risk of developing Alzheimer's disease and other forms of dementia than men. At the early stages of Alzheimer's disease, the most commonly recognised symptom is memory loss, especially for recently learned facts. Objects are misplaced or lost, and sufferers may repeatedly ask the same questions. The sufferer has difficulty in finding words to complete a sentence, and comprehension is poor, as is performance on complex motor tasks. There are also non-cognitive symptoms – delusions, depression, anxiety and verbal and physical aggression. Men, more than women, tend to become aggressive and to develop physical, verbal or sexual problems as the disease gets worse, while women become more reclusive and emotionally unstable. A longitudinal study on Swedish twins found that engaging in intellectual and cultural activities helps women to avoid the risk of developing Alzheimer's disease, but does not help men. Moreover, more women than men benefit from the protective effects of regular physical exercise.

Schizophrenia affects men and women to about the same extent, but women tend to be older at onset, and to suffer a more benign course of illness. The disorder in men peaks between the ages of fifteen and twenty-five years, whereas in women two peaks are seen, one at about thirty and another around menopause. Silvana Galderisi and her co-authors found that schizophrenic female patients, compared with male patients, showed fewer negative symptoms and less frequent alcohol abuse. But no significant difference was found between female and male patients in the rate of remission. Verbal memory is better preserved in women than in men. Female patients are more heavily paranoid, while male ones tend to display a subtype of the illness, characterised by disorganised behaviour and speech and inappropriate emotion.

The brains of healthy men and women differ in some ways from those of the mentally ill. Certain common differences between men and women are found to be reversed in mental illness. For instance, healthy men score higher than women in a task involving mental rotation of a three-dimensional image, as mentioned earlier, but schizophrenic women invariably outperform schizophrenic men when tested on this task. Differences in testosterone and oestrogen levels may explain these results, since in men and women who are mentally ill, as in schizophrenia, the women have higher levels of testosterone, while men have much lower levels than their healthy male counterparts. A history of childhood trauma is associated with a worse prognosis in male patients but not in female patients, or female controls.

Hypochondria is seen about equally in men and women, but this somatisation disorder characterised by recurring multiple clinically significant complaints about pain, stomach, sexual and neurological symptoms, for which no cause can be found, is more common in women. Women constitute the majority of chronic pain patients, and they are more sensitive to pain and differ in how they sense it and in their response to analgesics, so that they may require higher doses of morphine. Jenny Strong and her colleagues point out that women are also more willing to disclose pain, to complain of more symptoms and to be more emotional about them. Men's descriptions of pain are typically based on recounting facts and observations as well as their thoughts and emotions. Women tend to use more descriptive and evocative language, while men use fewer words and less graphic language, and take a more objective stance on their observations and recollections of the painful event. Women also suffer from chronic pain conditions such as headache and migraine more than men, and differences in pain threshold and tolerance have been demonstrated experimentally. That most patients with these conditions are female suggests that female hormones play an important role in the occurrence of the disorder. The finding that sixty per cent of women sufferers related attacks to their menstrual cycle supports this view. A placebo response to treatment was observed only in men.

But why are women more sensitive to pain than men? Women report pain more often and more severely than men in situations varying from recovering from surgery to arthritis. Experiments have been made where pain was induced to an equal degree in both men and women, and although there was similar activity in some brain regions of both sexes, several areas of male and female brains reacted differently. There was greater activity in the emotion-based centres of the women's brains, but in men it was the cognitive centres which showed greater activity. So maybe women's better empathy increases their perception of pain. Since women score higher than men on empathy, as we have seen, this could explain the findings that psychiatric disorders which are often characterised by a lack of empathy, such as antisocial behaviour, are far more common among boys and men.

Twice as many men as women are diagnosed with Parkinson's disease, the main symptoms of which are tremor, rigidity and slowness of movement. There is a link between Parkinson's disease and a loss of neurons using the chemical messenger dopamine in the substantia nigra of the mid-brain. A separate population of these neurons plays an important role in processes of reward and addiction, and sex differences have been seen in the development of dopaminergic neurons. The fact that women get Parkinson's later in life and less often than men may be explained by higher dopamine levels, possibly due to the activity of oestrogens. Men and boys are also more vulnerable to various other disorders with a major impact on movement such as Tourette's syndrome, which is characterised by multiple physical and verbal tics. It is three to four times more common in boys than in girls.

More than twice as many men as women become alcohol dependent during their lives. This may be due to women's physiology, which makes them more vulnerable than men. Even if they drink only the same amount of alcohol as men, women acquire higher concentrations of alcohol in their bloodstream and thus become drunk more quickly. They are also more likely than men to suffer organ damage. Many women may take this as a warning to limit how much alcohol they drink. There are more male drug abusers than female; for example, the male rate of marijuana smoking is twice the female rate.

Autism is a lifelong developmental disability that affects males more than females. It leads to difficulties in communication and relationships, and gives the individual severe problems in understanding the feelings and thinking of other people. It is a condition that begins before the child is three years old and has a strong genetic basis, and Cheslack-Postava and Jordan-Young have charted the changes. An infant with autism may not respond to their name, and when a little older may have delayed speech, and may fail to engage in pretend play with other children or may line up toys obsessively. An older child or an adult with autism may have difficulty understanding jokes, engaging in conversation and forming friendships. There is an impairment of mental skills in a substantial proportion of cases. The male incidence of autism is five to ten times higher than the female and is considered by some to be an extreme manifestation of the 'male brain', as males on average have a stronger drive to systemise, and autistic traits are often found in mathematicians, engineers, scientists, and their families. Such findings have led Simon Baron-Cohen and colleagues to suggest that high-functioning autism represents simply the high-systemising, low-empathising extreme of the population. There is a possible biological explanation for the disproportionate numbers of males who become autistic which is the effect of foetal testosterone, as there is a positive link between testosterone levels in the womb and the early signs of autism. There are brain differences too. The amygdala in males with autism is more enlarged than that of typical males early in development. As pointed out earlier, there are differences in brain functional connectivity which may also contribute to sex differences in diseases like autism, since the male brain has decreased local connectivity compared to the female.

However, a new theory has a different interpretation as to the cause of autism, and also the cause of psychotic disorders which are more common in women. Christopher Badcock has proposed that these disorders are based on the imprinted brain. As mentioned earlier, certain genes in the egg and sperm are imprinted during development and do not function. This new theory proposes that autism is due to a paternal bias in imprinted genes and that psychotic disorders are caused by a maternal bias in imprinted genes. Non-genetic factors such as nutrition in pregnancy can mimic and/or interact with imprinted gene expression, and the theory might even be able to explain the notable effect of maternal starvation on the risk of psychotic disorders as well as the 'autism epidemic' of modern affluent societies. Individuals with X-linked learning difficulties may also show this illness. An important example is Fragile X syndrome, expressed by forty-six per cent of males and sixteen per cent of females carrying a full mutation of a gene on the long arm of the X chromosome. Fragile X syndrome is the most common inherited cause of male intellectual disability and the best known single-gene cause of autism. It affects women rarely, and less severely.

Dyslexia impairs verbal fluency, comprehension, and reading ability and is found about twice as often in boys than girls between the ages of seven and fifteen. The greater variance of reading performance in males may account in part for their higher number. Women are more frequently prone to persistent vocal problems when speaking or singing, regardless of their occupation.

Many more boys than girls are affected by attention deficit and hyperactivity disorders. They are diagnosed twice to four times more often in boys than in girls. Attention deficit hyperactivity disorder – ADHD – is characterised by hyperactivity inattention and impulsivity. Overall, as Waddell and McCarthy have pointed out, males are more likely to suffer from disorders that occur early in development, such as hyperactivity disorders and learning disabilities, whereas females are more likely to develop mood disorders with later onset, such as depression. Critical periods of gonadal steroid release correspond to this difference in the onset of psychiatric disorders. It is, however, not yet possible to distinguish biological from societal and cultural influences on human brain development and behaviour. Waddell and McCarthy have also proposed, controversially, that males are at a higher risk for learning disabilities and hyperactivity because testosterone slows the early development of the brain, rendering males vulnerable to damage for longer periods of time. Aberrant developmental processes could occur during periods of dynamic embryological change in exposure to gonadal steroids. In males there is a perinatal sensitive period of elevated sex hormones that females do not experience, and this is a delicate period for brain development, when cellular processes like neuron differentiation and synapse formation could be affected.

Phobias, or irrational fears, take many forms and an estimated ten million people in the United Kingdom suffer from this anxiety disorder. Simple phobias can be the fear of things like spiders, snakes, enclosed spaces, flying, heights, injections or even going to the dentist. Simple phobias usually start early in life, but more complex phobias like agoraphobia can be life restricting and usually develop later. About twice as many women as men meet criteria for any specific phobia, and three times as many fear spiders or snakes. Phobias can have an evolutionary origin and non-human primates have evolved a fear mechanism specifically for snakes and spiders. This fear can be elicited in the first year of human life, especially in females, probably through women's historical exposure to these dangers while caring for their infants. Morbid fear of heights affects twice as many women as men.

Anorexia nervosa, an eating disorder where sufferers have an obsessive fear of gaining weight, is ten times more common in females, probably influenced by living in a culture where thinness is an ideal. Research has suggested that the fact that ten times more women get the disease than men and the fact that the most common time for it to start is at puberty may be related to an abnormal response of the brain to appetite-suppressing effects of the female sex hormone oestrogen. Abnormal oestrogen receptors are more commonly found in women with the disorder. Bulimia is characterised by cycles of eating too much in a short period of time – often in secret. It is also ten times more common in women. Individuals with bulimia alternate between bouts of excessive eating followed by periods of purging, with vomiting or the use of laxatives. The causes are not understood but again there may be a hormonal influence, since testosterone seems to protect men against eating disorders. Twin studies have shown that females who were in the womb with male twins have a lower risk for eating disorders than females with female twins, which suggests that testosterone from their male twin may protect them. Another illustration of sex difference is that exposure to severe malnutrition by children aged eleven to fourteen in the Dutch famine of 1944 made it much more likely that the females would develop diabetes and/or peripheral arterial disease at ages sixty to seventy-six, but that men did not have the same risk.

Sleep is another area where women may suffer more than men. Before puberty there are no significant differences between boys and girls, but adult women are twice as likely as men to have difficulties either falling asleep or staying asleep. One survey showed that seventy-eight per cent of women claimed not to have had a full night's sleep in the previous twelve months. Hormonal factors during pregnancy, lactation and menopause may lead to insomnia as well as psychological issues like depression or pain syndromes, which are both found more commonly in women.

Turning now to physical illnesses, cardiovascular disease is the most common cause of death in men and women worldwide. In the United Kingdom around one in six men and one in nine women die from it, says the British Heart Foundation. The incidence and the progression of cardiovascular disease and hypertension are much higher in men than in pre-menopausal women of the same age. But after menopause this is no longer true, and the incidence and rate of progression are similar. So men may develop most cardiovascular diseases, though not all of them, at an earlier age than women, but the number of affected women significantly increases with age.

Biological explanations for differences in the physical health of women and men often focus on the role of sex hormones. For example, oestrogen may protect pre-menopausal women from cardiovascular disease but may contribute to more women suffering from autoimmune disorders and pain conditions. It has been shown by Nadkarni and colleagues that this female sex hormone has an effect on white blood cells by moving a protein – annexin – from the surface of the cell into the interior, and thus preventing the cells from sticking to the insides of blood vessels, and causing dangerous blockages. Lower oestrogen levels after the menopause may be one reason why cardiovascular disease rates increase in women later in life and why these rates are higher throughout life in men. Oestrogen administration in postmenopausal women (HRT) has been associated with a significant reduction in the development of coronary artery disease and stroke. This has been interpreted as evidence that women's reproductive hormones give them protection them from these conditions. But the exact molecular mechanisms underlying this process have yet to be discovered and it may be that some differences are mediated by other mechanisms, especially by products of genes located on the X and Y chromosomes.

An important set of biological differences affecting health are sex-related differences in the nervous system's control of the heart. These may appear in measurements of heart-rate variability during the performance of a simple hand-grip motor task. Other differences may be seen in the relationship between regional blood flow and parasympathetic nervous activity which makes it possible for the body to recuperate and return to a balanced state. Females show an increased blood flow in the amygdala in response to parasympathetic activity whereas males show a reduction.

In heart failure risk factors and changes in heart muscle differ in men and women. A gene has been identified in one in five men, inherited from the father, which increases their risk of heart disease by fifty per cent. Women's heart muscle has been found to remodel itself more effectively after injury than that of men. This may be related to the sex hormones, that is, oestrogens and testosterone. Clinical analysis of the differences in disease of the heart valves supports this hypothesis. Clinical management in advanced countries differs between the sexes, with under-diagnosis and under-treatment typically being applied to women. But despite this, women frequently survive better than men. Men have a higher incidence of atrial fibrillation – an irregular heartbeat – than women. Lack of exercise is associated with an increased risk of heart disease in men, but surprisingly studies of women have produced mixed results. It seems that men who increase their level of activity can decrease their risk of heart disease, but this is not true for women. Diller, Patros and Prentice found differences in cardiovascular reactivity between the sexes in response to stress, as females tend to have greater heart-rate responses and males tend to have greater changes in blood pressure. As we have seen, women generally have better cardiac function and survival than men. However, it remains unclear whether there are sex differences in clinical features, treatment and prognosis after acute heart attacks. Some studies have reported no significant difference in mortality after adjusting for differences in age and other risk factors. Dreyer and her colleagues found that women with stable angina do worse than their male counterparts, and that several factors may contribute to this. These include clinical management, underlying biology and psychosocial issues.

Obesity is a growing problem, particularly in the United States and the United Kingdom, and both sexes carry risks. In 2012 in England twenty-five per cent of people, both men and women, were classified as obese by the NHS. But obese men may carry greater overall health risks, including risks for heart disease, than women. This is because more of them tend to carry the excess weight around their waist, rather than distributed in their hips and thighs, and are 'apple'- rather than 'pear'-shaped.

Autoimmune diseases include more than seventy different disorders caused by the antibodies whose normal role is to prevent infections from bacteria and viruses attacking the body. These diseases are overwhelmingly expressed in women, with over eighty per cent of patients being female. Very few autoimmune disorders show a male predominance. Although the evolutionary origin of the sexual immune difference is still unclear, it may be that women have a stronger immune system to save them from infections. This could be due to their role creating new life, thus requiring a stronger immune system to protect themselves so they can nurture the offspring. Type 1 diabetes, though an autoimmune illness, does not show an increased prevalence in women and is the only common organ-specific autoimmune disorder *not* to show a strong female bias.

Multiple sclerosis is a severe autoimmune illness affecting more than twice as many women as men. World-wide there are more than 2.3 million sufferers. It is caused by the sufferer's own antibodies destroying the cells that insulate the axons of neurons, causing a failure to transmit nerve impulses, and symptoms can occur in any part of the body. There are many different symptoms, the most common ones including numbness and tingling, problems with mobility and balance, muscle weakness and tightness and blurred vision.

Primary biliary cirrhosis is an autoimmune disease of the liver which has a nine-to-one female predominance. The main immunological characteristic of the illness is an antibody to mitochondria, the intracellular organelles which produce energy. This antibody cross-reacts with a product found in cosmetics, soaps and perfumes, and Lockshin has argued that lifestyle choices that are predominantly female, such as use of cosmetics and perfumes, lead to the development of this highly female-predominant disorder. The sex ratios of other autoimmune diseases are similarly more likely to be explained by environmental exposure than by intrinsic biological differences, he says.

Chronic fatigue syndrome, more common in women, is characterised by severe fatigue and headache, tender lymph nodes, joint and muscular pain, and an inability to concentrate. It may be related to the immune system, but this remains controversial. Women are also more likely than men to develop rheumatoid arthritis, a chronic systemic inflammatory disorder that principally attacks joints. The disease often leads to the destruction of cartilage as it progresses, and causes severe stiffness and pain in the joints. It is known that autoimmunity plays a key role in causing rheumatoid arthritis, but exactly what triggers the attack is unknown. About one per cent of the world's population is afflicted by rheumatoid arthritis, and for every man three women are sufferers. Recent data suggest that women also suffer greater disability than men with this disease. Women have stronger natural inflammatory responses than men when their immune systems are triggered. But although this may often result in superior immunity, if the immune system goes wrong it may also increase a woman's risk of developing an autoimmune disorder.

With stroke, a part of the brain dies from lack of blood. Globally stroke is more common among men, but women are more severely incapacitated. A factor causing stroke may be raised blood pressure, found to be higher in men than in women of similar ages. When men and women of the same age are compared, more men than women suffer strokes. But this difference vanishes in old age. Treatments also vary. Men are treated more often with intravenous thrombolysis, which causes the breakdown of blood clots by injection of pharmacological agents, compared with women. Aspirin seems more effective in reducing risk in men than in women, but statins have a similar effect in both sexes.

Cancer affects more men than women, and men are more likely than women to die from it. The two most common cancers are prostate cancer and breast cancer. The incidence of breast cancer in the United Kingdom is about 49,000 cases a year; for prostate about 41,000. The three cancers that affect only men are penile, prostate and testicular cancer. For men the three major killers are lung, prostate and bowel cancers, and for women, lung, breast and bowel. The incidence of lung cancer is greater in men than in women. In 2010 nearly 19,000 women were diagnosed with lung cancer in the UK, compared to 23,000 in men. Unlike men, the majority of women who develop lung cancer have never smoked.

For infectious diseases, in most settings, tuberculosis is more common in males at all ages except in childhood, when the reverse is the case. Only a third of the world's reported TB cases are in women. Malaria is equally likely to be contracted by men and women but pregnant women are at greater risk. The increased risk of seasonal and pandemic influenza to pregnant women and the elderly has been well documented. When women have a chronic cough they are more likely to go to the doctor than men, partly because they may also suffer from embarrassing complications such as stress incontinence. HIV is much more common among men owing both to their heterosexual and to their homosexual sexual activity.

Some sex differences in illness change with age. The common diseases of older men and women are similar, namely heart disease, cancer, musculoskeletal problems, diabetes, mental illness, impairments of sight and hearing, and incontinence. As we have seen men develop cardiovascular disease earlier than women. There are clear differences in health in old age between the sexes. Although women may have heart attacks later than men, they also tend to have a poorer quality of later life as a result of both physical limitations and depression. Osteoporosis, which is characterised by loss of bone mass leading to increased risk of fracture, affects around three million people in the United Kingdom. Women more often develop primary osteoporosis, which is related to the drop in oestrogen that occurs at menopause as a normal part of the ageing process.

According to brain-imaging studies, memory and attention deficits in ageing are related to a reduction of grey matter and to decreased activity in frontal and mediotemporal areas. There are no sex differences in the ageing of white matter, but the steepness of the trajectory along which grey matter volume decreases differs between men and women in almost all regions. There are greater age-related overall declines in brain volume in men than in women, possibly resulting in greater age-related overall losses of cognitive functioning. But findings are inconsistent. Several studies have reported older men to have outperformed their female controls in mental speed tasks. This result stands in contrast to results for memory, which favour women and illustrates the multi-dimensionality of cognition. However, these findings remain equivocal. There is also the suggestion that the best way to get a man to do something is to tell him he is too old for it!

One study in the United Kingdom carried out on 11,000 men and women aged over fifty and published as part of the Longitudinal Study on Ageing in 2010 found that women aged seventy-five and older reported a particularly low standard of well-being with many symptoms of depression, low life satisfaction, poor quality of life and high ratings of loneliness. In contrast men aged sixty-five and over seemed to be more satisfied with their lives than younger men. The researchers suggest that women affected by loneliness feel a greater sense of isolation when children leave home or when husbands, partners and friends are no longer around. Professor Sir Michael Marmot, lead researcher on the UK study, said: 'Older women are more likely to be living lives of loneliness because men die earlier.'

After their role in embryonic development, hormones continue to have important effects. Testosterone affects emotional tone and cognition. As women age, neurobehavioural functions such as sexual arousal or aggression decrease and the ratio of testosterone to oestrogen increases. For men, the reverse is true. Beginning at approximately fifty, men secrete progressively lower amounts of testosterone; about twenty per cent of men aged over sixty have lower than normal levels, and so the ratio of oestrogen to testosterone increases.

Although the neural mechanisms involved in ageing

have not been widely researched, it has been suggested that they could involve the same neural circuitry as that which is affected by hormones. A region of the suprachiasmatic nucleus, involved in the control of circadian rhythms which affect sleep patterns, contains twice as many neurons in men as in women until middle age, when the sex difference reverses and then ultimately disappears. Women live longer than men but only by an average of about four years, though the oldest person to have lived was a woman aged 125 while the oldest man was ten years younger at 115.

Changes in gene expression occur during ageing of the brain, affecting more than three times as many genes in men than in women. Women and men have equivalent numbers of up-regulated (more active) and down-regulated (less active) genes. In men there are reductions in volume in regions of the brain which are greater than those in women, especially in frontal and temporal lobes but also in whole brain volume. Brain volume reductions might lead to less ability to carry out particular tasks in men than in women. Women tend to show stronger reductions than men in the hippocampus and parietal lobe, possibly leading to more deficits in memory functions. But women still have better memories than men regardless of age. In both men and women the sex drive declines with age. This happens more quickly in the case of women, though their desire for sex does not seem to go altogether. The effects of the menopause are important, especially with the loss of sensitivity in the genital area that tends to occur at that time.

All of these examples go to show that sex differences in health and illness are significant.

13

Differences

> Well, there's a little bit of man in every woman and
> a little bit of woman in every man.
>
> Betty Smith

When I started this book I already knew that the differences
between men and women were a hotly disputed area. There
were many ingrained myths, for which the evidence is very
poor, such as women being less intelligent than men or
speaking much more than men, and some trivial myths for
which the evidence is quite good, such as men being better at
map-reading and women poorer at parallel parking. There
are many contrasting theories on the origin of differences
and whether they came from biological or social causes.
I knew little about the differences in emotions, skills and
brain structures. But having studied the evidence I am persuaded that there are significant biological differences that
affect how the two sexes behave. Further research, though, is
required to understand the neural basis of these differences.

Humans are very complex, their brains and social life
especially so, and so clear results from research are not
easy to come by. The evidence, however, is persuasive that
there are some fundamental biological differences between
men and women. Intellectual differences are small, but
differences in emotions are more significant, as are the
physical and physiological differences that arise during
development of the embryo and which have the most
profound effect.

These differences can be best understood in terms of Darwinian evolution, as they have made the two sexes better adapted to their environment and also, by having two sexes at all, have generated more variants for selection to act on. In spite of the importance of evolution in determining differences between males and females, it is often totally neglected in studies that deny the role of biologically based differences. It could even help male–female relationships if the fundamental biological differences between men and women were more widely understood and appreciated.

Evolution has resulted in men being modified women. The early development of the human embryo is similar in males and females, and is essentially female, with sexual differences appearing only at later stages. The default development of the embryo is female, and testosterone, together with some genes, plays a key role in causing male development, both in body and brain, so males are essentially modified females and that is why they have functionless breasts. Evolution cares only about reproduction and the survival of offspring and it has been crucial in determining the biological differences between men and women. Men were modified to fulfil tasks requiring speed and strength, like hunting and protection, and so in development grew to be significantly bigger and stronger than women. They are also more aggressive, and this can be a physical and social advantage enabling men to dominate women. Women can be aggressive too, but their aggression is rarely physical. Men are also more likely to take physical risks. These physical male advantages, together with women bearing and caring for children, are the main reasons why women have been discriminated against and subordinated for a very long time. The nature of sexual attraction today may still have echoes from our evolutionary past as signs of good health and a symmetrical body that is neither too fat nor too thin are considered important female characteristics in developed cultures.

Unlike men, women were selected to be loving carers for their children and this resulted in them having much stronger feelings of empathy. Women thus have an ability to share other people's feelings, to have a positive interest in other humans, and to excel in the ability to decode non-verbal emotional cues which plays an essential role in child care. When viewing sad humans they show enhanced brain activity in brain regions believed to be part of a mirror neuron system which supports empathy. Women are more emotional than men, show their emotions more and have better recall of emotional memories. They cry and smile more than men.

Men have less empathy than women and have evolved to be systemisers. Systemising is the intuitive ability and drive to analyse and construct systems which follow rules, and these systems often explore areas which are more abstract and not directly related to other humans. These differences between an empathetic brain and a systemising one can make male–female relationships a bit difficult at times.

Clear sex differences with a biological basis can be seen in the behaviour of children from a very early age and show the early expression of empathy and systemising. Newborn girls spend more time looking at faces, while boys are more interested in things like mechanical mobiles hung from the ceiling. Sex-typed toy preferences, such as boys preferring trains and cars, and girls preferring dolls, and other sex-typed differences in children's play and conversation would seem to be biologically determined and influenced by testosterone in the womb. However, social influences, such as how babies and children are treated by their parents, can play a role.

We know that there are genetically specified differences in male and female brains that can affect their behaviour, but there is still much to be learned. Development results in the adult male brain being ten per cent larger than the female brain. Females tend to have more localised regions of information-processing grey matter and more networking white matter relative to brain size than males, which may explain some intellectual differences. Although there is no difference in intelligence, males vary more widely in certain abilities – there are more males with both high and low scores when tested for maths – while women tend to be better at assimilating and integrating information, and at remembering it.

Several structures related to sexual behaviour are different in men and women, and there are also reports of structural differences in various other regions whose function remains to be further investigated. While significant brain differences are mainly due to development in the embryo, social factors must also play a role.

Reproduction is at the core of male–female differences. The evolutionary advantage of having two sexes is complex but probably depends on generating more differences in the offspring produced, so that only those that are best adapted are selected. The desire for children and hence for sex are genetically determined emotions in both men and women and were programmed into our genes in the course of evolution. A fundamental biological difference is that men are much more sexually motivated than women. Both during tactile genital stimulation and when viewing the same sexually arousing visual stimuli, the differences in brain activity are particularly marked. Male sexual orientation is based upon the direct effect of testosterone on the developing human brain, as shown by genetic disorders that prevent cells responding to testosterone. CAH causes excess testosterone in the womb and leads to an increased incidence of homosexual feelings in women, although most CAH women are heterosexual. Male homosexuality is influenced, surprisingly, by how many older brothers a male has.

There are numerous differences between men and women in their susceptibility to certain illnesses. Bearing children, with months of pregnancy and years of caring for offspring, has had far-reaching effects. Women's health in relation to infections was crucial and they evolved a very strong immune system which can unfortunately overact and cause autoimmune illnesses like multiple sclerosis. Women are more prone to suffer from chronic diseases such as depression and arthritis, while men are more likely to suffer cardiovascular disease, stroke and autism. These are clear examples of genetic origin.

There are no major intellectual differences between men and women. Possibly related to systemising is the major and well-established difference that men score higher than women on tests of mental rotation and on tests of spatial perception and orientation. Male hormones play a key role, and females with CAH have higher spatial ability than their sisters throughout life. Women generally score higher than men on memory tests based on finding objects in spatial displays. They can even recall objects that have been added, or have changed places, making them more efficient shoppers in a supermarket. Females have also been found to excel at landmark memory, that is, memory of images of differing types of structures. They are also better with episodic memory, that is long-term memory based on personal experiences, and with short-term memory.

The almost universal subjection of women by men is hard to explain in terms of the differences in the emotions and skills that I have discussed, but empathy possibly makes women more willing to do what men request. Much more significant is the greater physical strength and aggression of men, and the burden on women of childbirth and the bringing up of children, which have been genetically imposed.

A recent report by the European Union Agency for Fundamental Rights showed that in Europe one-third of women have been victims of violence and in Britain even more. In a different study it was found that more than half of women in Britain have experienced bullying or harassment at work in the past three years

Women do not talk more than men, but what they say can be very different. The difference between men and women in relation to language is either small or close to zero. Differences in language use thus probably have a strong social cause, although biological factors like empathy are also involved.

The absence of women from fields related to maths and technology is largely due to social and biological factors that result in them preferring not to work in these areas and their having to look after children, not to their abilities. It could also be that their greater empathy attracts them less to these subjects, or that they are discouraged by strong negative stereotypes of women being unsuitable to work in these fields. Better early education of girls would help to overcome these stereotypes and give them more confidence about entering such fields. It is important to get rid of negative female stereotypes in relation to maths, science and engineering, and there must be greater effort to support the care of children. There is evidence that women can do well in engineering if they are linked with similar colleagues and have the necessary training. This is a set of objectives to be strongly encouraged. It is also necessary to get rid of prejudice, as a recent study in the United States found that both male and female researchers tend to rate job applications from women for a post in science lower than those from equivalent men.

In an article by Kay and Shipman about their new book *The Confidence Code*, they claim that there is 'a vast confidence gap that separates the sexes. Compared with men, women don't consider themselves as ready for promotions, they predict they'll do worse on tests, and they generally underestimate their abilities. This disparity stems from factors ranging from upbringing to biology . . . The good news is that with work, confidence can be acquired. Which means that the confidence gap, in turn, can be closed.'

There is no doubt that biology, via evolution and genetics, has made men and women significantly different. But where skills are concerned, there are not so many differences between the sexes and serious efforts must be made to make this widely known, and for women to be urged to enter a wider variety of fields.

*

In writing this book, I have learned a good deal about myself and a great deal about women. For example, I knew virtually nothing about empathy. I try now to see if I really do lack empathy and make an effort to use it in relation to my family. But there is no good evidence that I am succeeding. I am a classic systemising male, after all.

References

1 Questions
Gray, J., *Men Are from Mars, Women Are from Venus: A Practical Guide for Improving Communication and Getting What You Want in Your Relationships*, Thorsons, 1993

2 Discrimination
de Beauvoir, S., *The Second Sex*, Vintage, 2010
Blundell, S., *Women in Ancient Greece*, Harvard, 1995
McElvaine, R. S., *Eve's Seed: Biology, the Sexes, and the Course of History*, McGraw-Hill, 2001

3 Modified Women
Ngun, T.C, N. Ghahrahmi, F. J. Sánchez, S. Bocklandt and E. Vilain, 'The Genetics of Sex Difference in Brain and Behaviours', *Frontiers in Neuroendocrinology* 32.2 (2011), 227–46.
Wolpert, L., *Developmental Biology: A Very Short Introduction*, Oxford University Press, 2011

4 Two Sexes
Brown S. K., N. C. Pedersen, S. Jafarishorijeh, D. L. Bannasch, K. D. Ahrens et al., 'Phylogenetic Distinctiveness of Middle Eastern and Southeast Asian Village Dog Y Chromosomes Illuminates Dog Origins', *PLoS ONE* 6.12 (2011), e28496
Dawkins, R., *The Selfish Gene*, Oxford University Press, 1976
Dixson, A. F., *Primate Sexuality*, Oxford University Press, 1998
Fedigan, L. M., *Primate Paradigms*, University of Chicago Press, 1992
Gomes, C. M., and C. Boesch, 'Wild Chimpanzees Exchange Meat for Sex on a Long-Term Basis', *PLoS ONE* 4.4 (2009), e5116
Hamilton, W. D., *Between Shoreham and Downe: seeking the key to natural beauty*. Inamori Foundation Kyoto prize Commemorative Lecture, 1993, reprinted in *Narrow Roads of*

Gene Land: The Collected Papers of W. D. Hamilton, vol. 3,
Last Words, ed. M. Ridley, Oxford University Press, 2005

Kauth, M. R. (ed.), *The Handbook of the Evolution of Human Sexuality,* Haworth, 2006

Long, J. A., 'Dawn of the Deed: the Origin of Sex', *Scientific American,* January 2011

Nettle, D., 'An evolutionary model of low mood states', *Journal of Theoretical Biology* 257 (2009), 100–3

Ridley, M., *The Red Queen: Sex and the Evolution of Human Nature,* Viking, 1993

Scheib, J. E., S. W. Gangestad and R. Thornhill, 'Facial attractiveness, symmetry and cues of good genes', *Proceedings of the Royal Society of London. Series B: Biological Sciences* 266 (1999), 1913–17

Strier, K. B., *Primate Behavioural Ecology,* Allyn and Bacon, 2007

Symons, D., *The Evolution of Human Sexuality,* Oxford, 1979

Valenzuela, N., 'Sexual Development and the Evolution of Sex Determination', *Sexual Development* 2 (2008), 64–72

Wolpert, L., 'Over the course of evolution, breasts became permanently enlarged to signal sexual receptivity', *The Independent,* 8 December 2004

5 Brain

Bao, A.-M., and D. F. Swaab, 'Sex Differences in the Brain, Behavior, and Neuropsychiatric Disorders', *Neuroscientist* 16 (2010), 188550–65

Byrne, R., 'Animal Cognition: Bring Me My Spear', *Current Biology* 17.5 (6 March 2007), R164–5

Cahill, L., 'His Brain, Her Brain', *Scientific American,* April 2005

Chou, K.-H., 'Sex-linked White Matter Microstructure of the Social and Analytic Brain', *NeuroImage* 54 (2011), 725–33

Fine, C., *Delusions of Gender: The Real Science Behind Sex Differences,* Icon, 2010

Gillies, G. E., and S. Arthur, 'Estrogen Actions in the Brain and the Basis for Differential Action in Men and Women: A Case for Sex-specific Medicines', *Pharmacological Reviews* 62 (2010), 155–98

Gong, G., E. He and A. C. Evans, 'Brain Connectivity: Gender Makes a Difference', *Neuroscientist* 17 (2011), 575–91

Groeschel, S., B. Vollmer, M. D. King and A. Connelly,

References

'Developmental Changes in Cerebral Grey and White Matter Volume from Infancy to Adulthood', *International Journal of Developmental Neuroscience* 28 (2001), 481–9

Haier, R. J., R. E. Jung, R. A. Yeo, K. Head and M. T. Alkire, 'The Neuroanatomy of General Intelligence: Sex Matters', *Neuroimage* 25.1 (March 2005), 320–7

Hamann, S., 'Sex Differences in the Responses of the Human Amygdala', *Neuroscientist* 11 (2005), 288–93

Hines, M., 'Sex-related Variation in Human Behaviour and the Brain', *Trends in Cognitive Sciences* 14 (2010), 448–56

Ingalhalikar, M., A. Smith, D. Parker, T. D. Satterthwaite, M. A. Elliott, K. Ruparel, H. Hakonarson, R. E. Gur, R. C. Gur and M. Verma, 'Sex Differences in the Structural Connectome of the Human Brain', *Proceedings of the National Academy of Sciences of the United States of America* (Jan. 14 2014) 111(2), 823–8

Jordan-Young, R., *Brainstorm: The Flaws in the Science of Sex Differences*, Harvard, 2010

Leonard, C. M., S. Towler, S. E. Welcome, L. K. Halderman, R. Otto, M. A. Eckert and C. Chiarello, 'Size Matters: Cerebral Volume Influences Sex Differences in Neuroanatomy', *Cerebral Cortex* 18 (2008), 2920–31

Luders, E., E. Gaser, K. L. Narr, and A. W. Toga, 'Why Sex Matters: Brain Size Independent Differences in Gray Matter Distributions between Men and Women', *Journal of Neuroscience*, 29 (2009), 14265–70

Luders, E., P. M. Thompson and A. W. Toga, 'The Development of the Corpus Callosum in the Healthy Human Brain', *Journal of Neuroscience* 30 (2010), 10985–90

Luders, E., and A. W. Toga, 'Sex Differences in Brain Anatomy', *Progress in Brain Research* 186 (2010), 3–12

Menzler, K., 'Men and Women Are Different: Diffusion Tensor Imaging Reveals Sexual Dimorphism in the Microstructure of the Thalamus, Corpus Callosum and Cingulum', *NeuroImage* 54 (2011), 2557–62

NeuroImage, Virtual Special Issue on Neuroimaging Gender Differences, 2012

Ngun, T. C., N. Ghahramani, F. J. Sánchez, S. Bocklandt and E. Vilain, 'The Genetics of Sex Differences in Brain and Behaviour', *Frontiers in Neuroendocrinology* 32 (2011), 227–46

References

Pfaff, D. W., *Man and Woman*, Oxford University Press, 2011

Savic, I. (ed.), *Sex Differences in the Human Brain: Their Underpinnings, and Implications*, Elsevier, 2010

Savic, I., A. Garcia-Falgueras and D. F. Swaab, 'Sexual Differentiation of the Human Brain in Relation to Gender Identity and Sexual Orientation', *Progress in Brain Research* 186 (2010), 41–62

Stevens, J. S., and S. Hamann, 'Sex Differences in Brain Activation to Emotional Stimuli: A Meta-analysis of Neuroimaging Studies', *Neuropsychologia* 50 (2012), 1578–93

Tomasi, D., and N. D. Volkow, 'Gender Differences in Brain Functional Connectivity Density', *Human Brain Mapping* 33 (2011), 849–60

Trabzuni, D., A. Ramasamy, S. Imran, R. Walker, C. Smith, M. E. Weale, J. Hardy and M. Ryten, 'Widespread sex differences in gene expression and splicing in the adult human brain', *Nature Communications* 22 (2013), 2771

Westerhausen, R., K. Kompus, M. Dramsdahl, L. E. Falkenberg, R. Grüne, H. Hjelmervik, K. Specht, K. Plessen and K. Hugdahl, 'A Critical Re-examination of Sexual Dimorphism in the Corpus Callosum Microstructure', *NeuroImage* 56 (2011), 874–80

Zaidi, Z. F., 'Gender Differences in Human Brain: a Review', *Open Anatomy Journal* 2 (2010), 37–55

6 Children

Ardila, A., M. Rosselli, A. Matute and O. Inozemtseva, 'Gender Differences in Cognitive Development', *Developmental Psychology* 47 (2011), 984–90

Benenson, J. F., R. Tennyson and R. W. Wrangham, 'Male More Than Female Infants Imitate Propulsive Motion', *Cognition* 21 (2011), 262–7

Berenbaum, S. A., and A. M. Beltz, 'Sexual Differentiation of Human Behaviour: Effects of Prenatal and Pubertal Organizational Hormones', *Frontiers in Neuroendocrinology* 32 (2011), 183–200

Cheslack-Postava, K. and R. M. Jordan-Young, 'Spectrum Disorders: Toward a Gendered Embodiment Model', *Social Science and Medicine* 74 (2012), 1667–74

Connellan, J., S. Baron-Cohen, S. Wheelwright, A. Batki and J. Ahluwalia, 'Sex Differences in Human Neonatal Social

Perception', *Infant Behavior and Development* 23 (2000), 113–18

Dedovic, K., M. Wadiwall, V. Engert and J. C. Pruessner, 'The Role of Sex and Gender Socialization in Stress Reactivity', *Developmental Psychology* 45 (2009), 45–55

Eliot, L., *Pink Brain, Blue Brain: How Small Differences Grow into Troublesome Gaps – and What We Can Do About It*, Oneworld, 2010

Hines, M., 'Sex-related Variation in Human Behaviour and the Brain', *Trends in Cognitive Sciences* 14 (2010), 448–56

Jadva, V., M. Hines and S. Golombok, 'Infants' Preferences for Toys, Colors, and Shapes: Sex Differences and Similarities', *Archives of Sexual Behavior* 39 (2010), 1261–73

Lippa, R. A., *Gender, Nature, and Nurture*, Psychology Press, 2nd edition, 2005

Martin, C. L., and D. N. Ruble, 'Patterns of Gender Development', *Annual Review of Psychology* 61 (2009), 353–81

Pasterski, V., M. E. Geffner, C. Brain, P. Hindmarsh, C. Brook, and M. Hines, 'Prenatal Hormones and Childhood Sex Segregation: Playmate and Play Style Preferences in Girls with Congenital Adrenal Hyperplasia', *Hormones and Behaviour* 59 (2011), 549–55

Richardson, H. L., 'Sleeping Like a Baby: Does Gender Influence Infant Arousability?', *Sleep* 33 (2010), 1055–60

Schumm, W. R., 'Re-evaluation of the "No Differences" Hypothesis Concerning Gay and Lesbian Parenting as Assessed in Eight Early (1979–1986) and Four Later (1997–1998) Dissertations', *Psychological Reports* 103 (2008), 275–30

Zakriski, A. L., J. C. Wright, and M. K. Underwood, 'Gender Similarities and Differences in Children's Social Behavior: Finding Personality in Contextualized Patterns of Adaptation', *Journal of Personality and Social Psychology* 88 (2005), 844–55

Zosuls, K. M., D. N. Ruble, C. S. Tamis-Lemonda, P. E. Shrout, M. H. Bornstein and F. K. Greulich, 'The Acquisition of Gender Labels in Infancy: Implications for Gender-typed Play', *Developmental Psychology* 45 (2009), 688–701

7 *Sex*

Bancroft, J., and C. A. Graham, 'The Varied Nature of Women's Sexuality: Unresolved Issues and a Theoretical Approach', *Hormones and Behaviour* 59 (2011), 717–29

Bao, A.-M., and D. F. Swaab, 'Sex Differences in the Brain, Behavior, and Neuropsychiatric Disorders', *Neuroscientist* 16 (2010), 550–65

Berenbaum, S. A., and A. M. Beltz, 'Sexual Differentiation of Human Behaviour: Effects of Prenatal and Pubertal Organizational Hormones', *Frontiers in Neuroendocrinology* 32 (2011), 183–200

Diamond, L. S., *Sexual Fluidity: Understanding Women's Love and Desire*, Harvard, 2008

Garcia-Falgueras, A., and D. F. Swaab, 'A Sex Difference in the Hypothalamic Uncinate Nucleus: Relationship to Gender Identity', *Brain* 131 (2008), 3132–46

Garcia-Falgueras, A., and D. F. Swaab, 'A Sex Difference in the Hypothalamic Uncinate Nucleus: Relationship to Gender Identity', *Brain* 131.12 (2008), 3132–6

Goode, S. D., *Paedophiles in Society: Reflecting on Sexuality, Abuse and Hope*, Palgrave Macmillan, 2011

Hamann, S., 'Sex Differences in the Responses of the Human Amygdala', *Neuroscientist* 11 (2005), 288–93

LeVay, S., *Gay, Straight, and the Reason Why: the Science of Sexual Orientation*, Oxford University Press, 2011

Lloyd, E. A., *The Case of the Female Orgasm: Bias in the Science of Evolution*, Harvard, 2005

Ngun, T. C., N. Ghahramani, F. J. Sánchez, S. Bocklandt and E. Vilain, 'The Genetics of Sex Differences in Brain and Behaviour', *Frontiers in Neuroendocrinology* 32 (2011), 227–46

Perrett, D., *In Your Face: The New Science of Human Attractiveness*, Palgrave Macmillan, 2010

Shanley, D. P., R. Sear R. Mace and T. B. Kirkwood, 'Testing evolutionary theories of menopause', *Proceedings of the Royal Society B* 274, no. 1628 (7 Dec. 2007) 2943–9

Swami, V., and M. J. Tovée, 'Female physical attractiveness in Britain and Malaysia. A Cross-cultural Study', *Body Image* 2 (2005) 115–28

Thornhill, R., and S. W. Gangestad, 'Facial Attractiveness', *Trends in Cognitive Sciences* 3 (1999), 452–60.

Wallen, K., and K. A. Lloyd, 'Female Sexual Arousal: Genital Anatomy and Orgasm in Intercourse', *Hormones and Behaviour* 59 (2011), 780–92

8 Emotions
Azim, E., D. Mobbs, B. Jo, V. Menon and A. L. Reiss, 'Sex Differences in Brain Activation Elicited by Humor', *Proceedings of the National Academy of Sciences of the USA* 102 (2005), 16496–501

Baillargeon, R. H., M. Zoccolillo, K. Keenan, S. Côté, D. Pérusse, H. Wu, M. Boivin and R. E. Tremblay, 'Gender Differences in Physical Aggression: a Prospective Population-based Survey of Children Before and After Two Years of Age', *Developmental Psychology* 43 (2007), 13–26

Baron-Cohen, S., *The Essential Difference: Men, Women and the Extreme Male Brain*, Penguin, 2003

Baron-Cohen, S., and S. Wheelwright, 'The Empathy Quotient: An Investigation of Adults with Asperger Syndrome or High Functioning Autism,and Normal Sex Differences,' *Journal of Autism and Developmental Disorders* 34, no. 2 (2004), 163–75

Bianchin, M., and A. Angrilli, 'Gender Differences in Emotional Responses: A Psychophysiological Study', *Physiology and Behavior* 105 (2011), 925–32

Böhnke, R., K. Bertsch, M. R. Kruk, S. Richter and E. Naumann, 'Exogenous Cortisol Enhances Aggressive Behaviour in Females, but Not in Males', *Psychoneuroendocrinology* 35 (2010), 1034–44

Chapman, E., S. Baron-Cohen, B. Auyeung, R. Knickmeyer, K. Taylor and G. Hackett, 'Fetal Testosterone and Empathy: Evidence from the Empathy Quotient (EQ) and the "Reading the Mind in the Eyes" Test', *SocialNeuroscience* 1 (2006), 135–48

Cross, C. P., L. T. Copping and A. Campbell, 'Sex Differences in Impulsivity: A Meta-analysis', *Psychological Bulletin* 137 (2011), 97–130

Del Giudice, M., T. Booth and P. Irwing, 'The Distance Between Mars and Venus: Measuring Global Sex Differences in Personality', *PLoS ONE* 7.1 (2012), e29265

Derntl, B., A. Finkelmeyer, S. Eickhoff, T. Kellermann, D. I.

Falkenberg, F. Schneider and U. Habel, 'Multidimensional
Assessment of Empathic Abilities: Neural Correlates and
Gender Differences', *Psychoneuroendocrinology* 35 (2010),
67–82

de Vignemont, F., and T. Singer, 'The Empathic Brain: How, When
and Why', *Trends in Cognitive Sciences* 10 (2006), 435–41

Geary, D. C., *Male, Female: the Evolution of Human Sex
Differences*, American Psychological Society, 1998

Ginsburg, J., 'Sex Differences in Children's Physical Risktaking
Behaviors: Natural Observations at the San Antonio
Zoological Gardens', *North American Journal of Psychology* 9
(2007), 407–14

Goldstein, J. M., M. Jerram, B. Abbs, S. Whitfield-Gabrieli and
N. Makris, 'Sex Differences in Stress Response Circuitry
Activation Dependent on Female Hormonal Cycle', *Journal of
Neuroscience* 30 (2010), 431–8

Hamann, S., 'Sex Differences in the Responses of the Human
Amygdala', *Neuroscientist* 11 (2005), 288–93

Hertenstein, M. J., and D. Keltner, 'Gender and the
Communication of Emotion Via Touch', *Sex Roles* 64 (2011),
70–80

Knickmeyer, R., S. Baron-Cohen, P. Raggatt and K. Taylor, 'Foetal
testosterone, social relationships and restricted interests in
children', *Journal of Child Psychology and Psychiatry* 46:2
(2005) 198–210

Kohn, N., T. Kellermann, R. C. Gur, F. Schneider and U. Habel,
'Gender Differences in the Neural Correlates of Humor
Processing: Implications for Different Processing Modes',
Neuropsychologia 49 (2011), 888–97

Kret, M. E., S. Pichon, J. Grèzes and B. de Gelder, 'Men Fear
Other Men Most: Gender Specific Brain Activations in
Perceiving Threat from Dynamic Faces and Bodies – an fMRI
Study', *Frontiers in Emotion Science* 2 (2011), 1–11

Kuschel, R., ' "Women Are Women and Men Are Men": How
Bellonese Women Get Even', in K. Bjorkqvist and P. Niemele
(eds), *Of Mice and Women: Aspects of Female Aggression*,
Academic Press, 1992, 173–85

Lawrence, E. J., P. Shaw, D. Baker, S. Baron-Cohen and A. S.
David, 'Measuring Empathy: Reliability and Validity of the
Empathy Quotient', *Psychological Medicine* 34 (2004), 911–24

References

Levy, K. N. and K. M. Kelly, 'Sex differences in jealousy:
a contribution from attachment theory', 21:2 (2010)
Psychological Science, 168–73

MacCoby, E., *The Two Sexes: Growing Up Apart, Coming
Together*, Harvard University Press, 1998

Mercadillo, R. E., J. L. Díaz, E. H. Pasaye and F. A. Barrios,
'Perception of Suffering and Compassion Experience: Brain
Gender Disparities', *Brain and Cognition* 76 (2011), 5–14

Mestre, M. V., P. Samper, M. D. Frías and A. M. Tur, 'Are Women
More Empathetic Than Men? A Longitudinal Study in
Adolescence', *Spanish Journal of Psychology* 12 (2009), 76–83

Nettle, D., 'Empathizing and Systemizing: What Are They,
and What Do They Contribute to Our Understanding of
Psychological Sex Differences?'*British Journal of Psychology*
98 (2007), 237–55

Ohrmann, P., A. Pedersen, M. Braun, J. Bauer, H. Kugel, A.
Kersting, K. Domschke, J. Deckert and T. Suslow, 'Effect of
Gender on Processing Threat-related Stimuli in Patients with
Panic Disorder: Sex Does Matter' *Depress Anxiety* 27 (2010),
1034–43

Proverbio, A. M., R. Adorni, A. Zani and L. Trestianu, 'Sex
Differences in the Brain Response to Affective Scenes With or
Without Humans', *Neuropsychologia* 47 (2009), 2374–88

Provine, R., *Laughter: A Scientific Investigation*, Penguin, 2009

Schulte-Rüther, M., H. J. Markowitsch, N. J. Shah, G. R. Fink
and M. Piefke, 'Gender Differences in Brain Networks
Supporting Empathy', *NeuroImage* 42 (2008), 393–403

Van Honk, J., D. J. Schutter, P. A. Bos, A. W. Kruijt, E. G. Lentjes
and S. Baron-Cohen, 'Testosterone Administration Impairs
Cognitive Empathy in Women Depending on Second-to-
Fourth Digit Ratio', *Proceedings of the National Academy of
Sciences of the USA* 108 (2011), 3448–52

Vigil, J. M., 'A Socio-relational Framework of Sex Differences in
the Expression of Emotion', *Behavioural and Brain Sciences*
32 (2009), 375–428

9 *Mathematics*

Beilock, S. L., E. A. Gunderson, G. Ramirez and S. C. Levine,
'Female Teachers' Math Anxiety Affects Girls' Math
Achievement', *Proceedings of the National Academy of*

Sciences of the United States of America 107 (2010), 1860–3

Ceci, S. J., and W. M. Williams, 'Sex Differences in Maths-Intensive Fields', *Current Directions in Psychological Science* 19 (2010), 275–9

Ceci, S. J., and W. M. Williams, 'Understanding Current Causes of Women's Underrepresentation in Science', *Proceedings of the National Academy of Sciences of the USA* 108 (2011), 3157–62

Chen, J. C., S. Owusu-Ofori, D. Pai, E. Toca-McDowell, S.-L. Wang and C. K. A. Waters, 'A Study of Female Academic Performance in Mechanical Engineering', *Proceedings of the Frontiers in Education 26th Annual Conference* vol. 2 (1996), 779–82

Flory, J. A., A. Leibbrandt and J. A. List, *Do Competitive Work Places Deter Female Workers?* National Bureau of Economic Research, 2010

Halpern, D. F., C. F. Benbow, D. C. Geary, R. C. Gur, J. S. Hyde and M. A. Gernsbacher, 'The Science of Sex Differences in Science and Maths', *Psychological Science in the Public Interest* 8 (2007), 1–51

Haworth, C. M., P. S. Dale and R. Plomin, 'Sex Differences in School Science Performance from Middle Childhood to Early Adolescence', *International Journal of Educational Research* 49 (2010), 92–101

Lawrence, P. A., 'Men, Women, and Ghosts in Science', *PLoS Biol* 4.1 (2006), e19

Whaley, L. A., *Women's History as Scientists: A Guide to the Debates*, ABC-Clio, 2003

10 Skills

Berenbaum, S. A., K. L. Bryk and A. M. Beltz, 'Early Androgen Effects on Spatial and Mechanical Abilities: Evidence from Congenital Adrenal Hyperplasia', *Behavioral Neuroscience* 126 (2012), 86–96

Cohen, N. R., M. Pomplun, B. J. Gold and R. Sekuler, , 'Sex Differences in the Acquisition of Complex Skilled Movements', *Experimental Brain Research* 205 (2010), 183–93

Hahn, N., P. Jansen and M. Heil, 'Preschoolers' Mental Rotation: Sex Differences in Hemispheric Asymmetry', *Journal of Cognitive Neuroscience* 22 (2010), 1244–50

References

Hassan, B., and Q. Rahman, 'Selective Sexual Orientation-related Differences in Object Location Memory', *Behavioral Neuroscience* 121 (2007), 625–33

Hausmann, M., D. Slabbekoorn, S. H. Van Goozen, P. T. Cohen-Kettenis and O. Güntürkün, 'Sex hormones affect spatial abilities during the menstrual cycle,' *Behavioral Neuroscience* 114 (2000), 1245–50

Hines, M., B. A. Fane, V. L. Pasterski, G. A. Mathews, G. S. Conway and C. Brook, 'Spatial Abilities Following Prenatal Androgen Abnormality: Targeting and Mental Rotations Performance in Individuals with Congenital Adrenal Hyperplasia', *Psychoneuroendocrinology* 28 (2003), 1010–26

Hines, M., 'Sex-related Variation in Human Behaviour and the Brain', 194 *Trends in Cognitive Sciences* 14 (2010), 448–56

Hyde, J. S., 'The Gender Similarities Hypothesis', *American Psychologist* 60 (2005), 581–92

Lippa, R. A., M. L. Collaer and M. Peters, 'Sex Differences in Mental Rotation and Line Angle Judgments Are Positively Associated with Gender Equality and Economic Development Across Fifty-three Nations', *Archives of Sexual Behavior* 39 (2010), 990–7

Maylor, E. A., S. Reimers, J. Choi, M. L. Collaer, M. Peters and I. Silverman, 'Gender and Sexual Orientation Differences in Cognition Across Adulthood: Age Is Kinder to Women Than to Men Regardless of Sexual Orientation', *Archives of Sexual Behaviour* 36:2 (2007) 235–49

Mendrek, A., N. Lakis and J. Jiménez, 'Associations of sex steroid hormones with cerebral activations during mental rotation in men and women with schizophrenia', *Psychoneuroendocrinology*, 36 (2011), 1422–6

Rahman, Q., C. Newland and B. M. Smyth, 'Sexual Orientation and Spatial Position Effects on Selective Forms of Object Location Memory', *Brain and Cognition* 75 (2011), 217–24

Ruggieroa, G., I. Sergi and T. Iachini, 'Gender Differences in Remembering and Inferring Spatial Distances', *Memory* 16.8 (2008), 821–35

Wraga, M., M. Helt, E. Jacobs and K. Sullivan, 'Neural basis of stereotype-induced shifts in women's mental rotation performance', *Social Cognitive and Affective Neuroscience* 2.1 (2007), 12–19

Zündorf, I. C., H.-O. Karnath and J. Lewald, 'Male Advantage in Sound Localization at Cocktail Parties', *Cortex* (2011), 741–9

11 Language

Allendorfer, J. B., C. J. Lindsell, M. Siegel, C. L. Banks, J. Vannest, S. K. Holland and J. P. Szaflarski, ' Females and males are highly similar in language performance and cortical activation patterns during verb generation', *Cortex* (2012), 1218–33

Barnes, D., 'Language and Learning in the Classroom', *Journal of Curriculum Studies* 3:1 (1971), 27–38

Burman, D. D., 'Sex Differences in Neural Processing of Language among Children', *Neuropsychologia* 46 (2008), 349–62

Burman, D. D., T. Bitan and J. R. Booth, 'Sex Differences in Neural Processing of Language among Children,' *Neuropsychologia* 46 (2008), 1349–62

Cameron, D., *The Myth of Mars and Venus*, Oxford University Press, 2007

Chiarello, C., S. E. Welcome, L. K. Halderman, S. Towler, J. Julagay, R. Otto and C. M. Leonard, 'A Large-scale Investigation of Lateralization in Cortical Anatomy and Word Reading: Are There Sex Differences?', *Neuropsychology* 23 (2009), 210–22

Diprete, T. A., and J. L. Jennings, 'Social and Behavioral Skills and the Gender Gap in Early Educational Achievement', *Social Science Research* 41 (2011) 1–15

Galsworthy, M. J., G. Dionne, P. S. Dale, and R. Plomin, 'Sex Differences in Early Verbal and Non-verbal Cognitive Development', *Developmental Science* 3 (2000), 206–15

Gauthier, C. T., M. Duyme, M. Zanca and C. Capron, 'Sex and Performance Level Effects on Brain Activation During a Verbal Fluency Task: A Functional Magnetic Resonance Imaging Study', *Cortex* (2009) 164–76

Havy, M., J. Bertoncini and T. Nazzi, 'Word Learning and Phonetic Processing in Preschool-age Children', *Journal of Experimental Child Psychology* 108 (2011), 25–43

Hawke, J. L., R. K. Olson, E. G. Willcutt, S. J. Wadsworth and J. DeFries, 'Gender Ratios for Reading Difficulties', *Dyslexia* 15 (2009), 239–42

Hyde, J., and M. Linn, 'Gender Differences in Verbal Ability: A Meta-analysis', *Psychological Bulletin* 104.1 (1988), 53–69

References

Kaushanskaya, M., V. Marian and J. Yoo, 'Gender Differences in Adult Word Learning', *Acta Psychologica* 137 (2011), 24–35

Logan, S., and R. Johnston, 'Investigating Gender Differences in Reading', *Educational Review* 62 (2010), 175–87

Mehl, M. R., S. Vazire, N. Ramírez-Esparza, R. B. Slatcher and J. W. Pennebaker, 'Are Women Really More Talkative Than Men?', *Science* 317 (2007), 82

Newman, M. L., C. J. Groom, L. D. Handelman and J. W. Pennebaker, 'Gender Differences in Language Use: an Analysis of 14,000 Text Samples', *Discourse Processes* 45 (2008), 211–36

Ozçalişkan, S., and S. Goldin-Meadow, 'Sex Differences in Language First Appear in Gesture', *Developmental Science* 13 (2010), 752–60

Sommer, I. E., A. Aleman, M. Somers, M. P. Boks and R. S. Kahn, 'Sex Differences in Handedness, Asymmetry of the Planum Temporale and Functional Language Lateralization', *Brain Research* 1206 (2008), 76–88

Tannen, D. 'Gender and Family Interaction', in J. Holmes and M. Meyerhoff, *The Handbook on Language and Gender* Basil Blackwell, 2013, 179–201

Tenenbaum, H. R., S. Ford and B. Alkhedairy, 'Telling Stories: Gender Differences in Peers' Emotion Talk and Communication Style', *British Journal of Developmental Psychology* 29 (2011), 707–21

Van den Brink, D., J. J. Van Berkum, M. C. Bastiaansen, C. M. Tesink, M. Kos, J. K. Buitelaar and P. Hagoort, 'Empathy Matters: ERP Evidence for Inter-individual Differences in Social Language Processing', *Neuropsychologia* 46 (2008), 1349–62

Vigil, J. M., 'A Socio-relational Framework of Sex Differences in the Expression of Emotion', *Behavioural and Brain Sciences* 32 (2009), 375–90

Wallentin, M., 'Putative Sex Differences in Verbal Abilities and Language Cortex: A Critical Review', *Brain and Language* 108 (2009), 175–8

Zimmerman, D., and C. West, 'Sex Roles, Interruptions and Silences in Conversation', in B. Thorne and N. Henley (eds), *Language and Sex: Difference and Dominance*, Newbury House, 1977, 105–29

12 Health

Badcock, C., 'The Imprinted Brain: How Genes Set the Balance between Autism and Psychosis', *Epigenomics* 3 (2011), 345–59

Cheslack-Postava, K., and R. M. Jordan-Young, 'Autism Spectrum Disorders: Toward a Gendered Embodiment Model,' *Social Science and Medicine* 74 (2012), 1667–74

Diller, J. W., C. H. Patros and P. R. Prentice, 'Temporal Discounting and Heart Rate Reactivity to Stress', *Behavioural Processes* 87 (2011), 306–9

Dreyer, R., M. Arstall, R. Tavella, C. Morgan, A. Weekes and J. Beltram, 'Gender Differences in Patients with Stable Angina Attending Primary Care Practices', *Heart, Lung and Circulation* 20 (2011), 452–9

Galderisi, S., P. Bucci, A. Üçok, and J. Peuskens, 'No Gender Differences in Social Outcome in Patients Suffering from Schizophrenia', *European Psychiatry* 27 (2012), 406–8

Guarner-Lans, V., M. E. Rubio-Ruiz, I. Pérez-Torres and G. Baños de MacCarthy, 'Relation of Aging and Sex Hormones to Metabolic Syndrome and Cardiovascular Disease', *Experimental Gerontology* 46 (2011), 517–23

Kay, K., and C. Shipman, *The Confidence Code: The Science and Art of Self-Assurance – What Women Should Know* HarperCollins, 2014

Kryspin-Exner, I., E. Lamplmayr and A. Felnhofer, 'Geropsychology: The Gender Gap in Human Aging – a Mini-review', *Gerontology* 57 (2011), 539–48

Lockshin, M. D., 'Nonhormonal Explanations for Sex Discrepancy in Human Illness', *Annals of the New York Academy of Sciences* 193 (2010), 22–4

Maguen, S., L. Ren, J. O. Bosch, C. R. Marmar and K. H. Seal, 'Gender Differences in Mental Health Diagnoses among Iraq and Afghanistan Veterans Enrolled in Veterans Affairs Health Care', *American Journal of Public Health* 100 (2010), 2450–6

Mittelstrass, K., J. S. Ried, Z. Yu, J. Krumsiek, C. Gieger, C. Prehn, W. Roemisch-Margl, A. Polonikov, A. Peters, F. J. Theis, T. F. Meitinger, S. Weidinger, H. E. Wichmann, K. Suhre, R. Wang-Sattler, J. Adamski and T. Illig, 'Discovery of Sexual Dimorphisms in Metabolic and Genetic Biomarkers', *PLoS Genetics* 7 (2011), e1002215

References

Nadkarni, S., D. Cooper, V. Brancaleone, S. Bena and M. Perretti, 'Activation of the Annexin A1 Pathway Underlies the Protective Effects Exerted by Estrogen in Polymorphonuclear Leukocytes', *Arteriosclerosis, Thrombosis, and Vascular Biology* 31 (2011), 2749–59

Nasser, E. H., N. Walders and J. H. Jenkins, 'The Experience of Schizophrenia: What's Gender Got to Do with It? A Critical Review of the Current Status of Research on Schizophrenia', *Schizophrenia Bulletin* 28 (2002), 351–62

Needham, B., and T. D. Hill, 'Do Gender Differences in Mental Health Contribute to Gender Differences in Physical Health?', *Social Science and Medicine* 71 (2010), 1472–9

Piccinelli, M., and G. Wilkinson, 'Gender Differences in Depression: Critical Review', *British Journal of Psychiatry* 177 (2000), 486–92

Rakison, D. H., 'Does Women's Greater Fear of Snakes and Spiders Originate in Infancy?', *Evolution and Human Behaviour* 30 (2009), 439–44

Rogers, N., M. Stafford and A. Steptoe, *Financial Circumstances, Health and Well-Being of the Older Population in England: The 2008 English Longitudinal Study of Ageing* (Wave 4), Institute for Fiscal Studies, 2010

Strong, J., T. Mathews, R. Sussex, F. New, S. Hoey, and G. Mitchell, 'Pain Language and Gender Differences When Describing a Past Pain Event', *Pain* 145 (2009), 86–95

Van Abeelen, A. F., S. G. Elias, P. M. Bossuyt, D. E. Grobbee, Y. T. Van der Schouw, T. J. Roseboom and C. S. Uiterwaal, 'Famine Exposure in the Young and the Risk of Type 2 Diabetes in Adulthood' *Diabetes* 61:9 (2012) 2255–60

Waddell, J. and M. M. McCarthy, 'Sexual Differentiation of the Brain and ADHD: What Is a Sex Difference in Prevalence Telling Us?', *Current Topics in Neuroscience* 9 (2012), 341–60

Index